Conspiracy Theories

Recent Releases from Open Court

Spoiler Alert! (It's a Book about the Philosophy of Spoilers)
Richard Greene

Mister Rogers and Philosophy: Wondering through the Neighborhood
Edited by Eric J. Mohr and Holly K. Mohr

RuPaul's Drag Race and Philosophy: Sissy That Thought
Edited by Hendrik Kempt and Megan Volpert

Scott Adams and Philosophy: A Hole in the Fabric of Reality
Edited by Daniel Yim, Galen Foresman, and Robert Arp

The Handmaid's Tale and Philosophy: A Womb of One's Own
Edited by Rachel Robison-Greene

WikiLeaking: The Ethics of Secrecy and Exposure
Edited by Christian Cotton and Robert Arp

For full details of all Open Court books, visit www.opencourtbooks.com.

Conspiracy Theories

Philosophers Connect the Dots

Edited By
RICHARD GREENE
and
RACHEL ROBISON-GREENE

OPEN COURT
Chicago

To find out more about Open Court books, visit our website at www.opencourtbooks.com.

Open Court Publishing Company is a division of Carus Publishing Company, dba Cricket Media.

Copyright © 2020 by Carus Publishing Company, dba Cricket Media

First printing 2020

All rights reserved. No part of this publication may be reproduced, stored in a retrieval system, or transmitted, in any form or by any means, electronic, mechanical, photocopying, recording, or otherwise, without the prior written permission of the publisher, Open Court Publishing Company, 70 East Lake Street, Suite 800, Chicago, Illinois 60601.

Printed and bound in the United States of America.

Conspiracy Theories: Philosophers Connect the Dots

ISBN: 978-0-8126-9479-6

Library of Congress Control Number: 2019953286

This book is also available as an e-book.

For Clint

Contents

Thanks

Working on this project has been a pleasure, in no small part because of the many fine folks who have assisted us along the way.

In particular, a debt of gratitude is owed to David Ramsay Steele at Open Court, the contributors to this volume, and our respective academic departments at Utah State University and Weber State University.

We'd also like to thank those family members, students, friends, and colleagues with whom we've had fruitful and rewarding conversations on various aspects of all things conspiratorial as they relate to philosophical themes.

We are very grateful to the Pierogi Gallery and the Lombardi family for permission to reproduce art by Mark Lombardi in Chapter 19. Mark Lombardi, "George W. Bush, Harken Energy, and Jackson Stevens, c. 1979–90 (5th Version)." 1999, Graphite on paper, 20 x 44 inches. Image courtesy Pierogi Gallery and The Lombardi Family. Photo Credit: John Berens. Private Collection.

If Only There Were a Vaccine against Gullibility

On the one hand, conspiracy theories can be a lot of fun, and in many cases are pretty innocuous. Consider, for example, the recent "birds aren't real" movement. Almost no one buys into this one, and all things considered, it's pretty funny. Certain other conspiracy theories, such as "The moon landing was faked," have believers, but their beliefs have virtually no effect on the affairs of others. Who cares if some people believe the Earth is flat, or that the Egyptian pyramids were built by aliens?

On the other hand, conspiracy theories can lead to real problems. This is because they obscure the truth, which can lead to pretty severe harms. For example, anti-vaxxers have contributed greatly to the return of diseases that had been virtually eradicated for decades, birthers have exacerbated racial tensions, climate skeptics support policies that are not consistent with the long-term survival of our species. Moreover, the fact that conspiracy theories are fun amplifies the problems that arise when they are not innocuous. This is because conspiracy theories are not just propagated by true believers; some people will spread conspiracies that they don't believe, just because it is fun or cool to do so. Frankly, it's trendy in certain circles to believe in the Illuminati or to believe that there are aliens being held in Area 51.

Given that the stakes surrounding conspiracy theories are sometimes pretty high, some serious sorting out is required. Right off the bat, we need clarification on what conspiracy theories are, how they function, when they are harmful, precisely how they are harmful, and when they are just amusing.

Moreover, we need to take a close look at the particulars of some of the most enduring conspiracy theories. That's precisely what the philosophers in this book are doing. The payoff will be a roadmap for distinguishing good theories from bad ones, and harmful ones from innocuous ones, while exposing some of the worst conspiracy theories. We may never know whether birds are actually real, but we will know whether we can throw away our tinfoil hats.

PART I

---

*"If only you assume
a big enough conspiracy,
you can explain
anything, including the
cosmos itself."*

1
From Alien Shape-shifting Lizards to the Dodgy Dossier

M R.X. DENTITH

Who comes to mind when you think about conspiracy theorists? If you happened to watch blockbusters back in the late Nineties, you might think of Jerry Fletcher from the movie *Conspiracy Theory*, or Edward "Brill" Lyle from *Enemy of the State*. They were conspiracy theorists who believed that shadowy members of secret groups were directing world affairs.

While Fletcher and Brill turned out to be correct, they were really only right by accident: in *Conspiracy Theory*, for example, the villain of the piece (played by Patrick Stewart with some approximation of an American accent) revealed that they have been feeding Fletcher's paranoid tendencies and helping his investigations in order to help ferret out threats to the conspirators' nefarious plans.

Conspiracy Theory and *Enemy of the State* played upon the notion of the paranoid conspiracy theorist. What made them believable was the feeling that they were drawn from real life. Indeed, often when we think of conspiracy theorists in the real world we tend to think of people like Alex Jones or David Icke.

For the uninitiated, Alex Jones runs InfoWars, a website famous for its conspiratorial content. Jones believes a lot of conspiracy theories: mass shootings are really false flag operations being run by the government to bring in draconian gun control measures; Big Pharma is turning frogs gay; Michelle Obama is secretly a trans woman); and FEMA is building death camps in the middle of the US to prepare for the New World Order takeover.

David Icke, from across the Pond, is a former British footballer turned BBC sports journalist turned messianic figure, who has found another second life touring the world giving

eight-hour lectures on how we are secretly controlled by alien
shape-shifting lizards. His theories encompass everything,
from recurrent symbols in world religions, George Soros's total
control of the liberal agenda, and the Moon being an artificial
and hollow satellite broadcasting a signal originating from
Saturn which is trapping us in a hologrammatic prison. Like
Jones, he is celebrated by some . . . and considered by others to
be a tinfoil-hat-wearing weirdo who believes any conspiracy
theory he hears.

But are Jones and Icke really typical examples of conspiracy theorists, or are they just notable? After all, someone can
be famous without actually being representative of the group
they belong to. Donald J. Trump is a notable Republican, but
many people would like to think he isn't a typical member of
the US Right. Similarly, we intuitively might think of conspiracy theorists as being like the Jerry Fletchers or the
David Ickes of this world. But unless we understand what it
is to be a conspiracy theorist, that intuition could turn out to
be misguided.

What Exactly Is a Conspiracy Theory?

In order to understand who conspiracy theorists are, we need
to do a little philosophical table-setting. Philosophers like to
work with precisely defined concepts, and so we need to say
something about what counts as a "conspiracy theory" before
we can say who counts as a "conspiracy theorist."

Luckily for us, there has been—over the last two decades—
quite a lot of scholarly discussion on precisely this matter.

Most conspiracy theory theorists (the scholars who theorize
about conspiracy theories) agree that a "conspiracy" is simply

1. **a plan between two or more people who**

2. **work in secret**

3. **towards some end.**

You need all three of these things to be true to be part of a
conspiracy (you can't conspire alone), but that's all you need
(if you're doing all three things you are conspiring!). (If you
want more detail on how scholars got to this conclusion, you
can read the chapter I wrote for *Conspiracy Theories and the
People Who Believe Them*, edited by Joseph Uscinksi).

A conspiracy theory then, is just a theory about a conspiracy, right? So, there shouldn't be anything particularly suspi-

cious about believing in them. After all, we know conspiracies occur, and theorizing about conspiracies surely isn't off-limits. So, what's wrong with being a conspiracy theorist, someone who simply theorizes about conspiracies?

Suspicious Conspiracy Theories

"Ah," but, you might be thinking, "this is not what we normally mean when we say 'conspiracy theory'." That is, you might think that ordinarily—rather than the kind of jargon philosophers use—a conspiracy theory is more than just a theory about a conspiracy. Rather, when we talk about conspiracy theories, we're talking about something which is definitely implied to be suspicious!

That assumption might not survive scrutiny, however. Here's an example as to why. In the lead up to the 2014 General Election in Aotearoa/New Zealand, then Prime Minister John Key tried to deny that his office was engaging in dirty politics by claiming such stories were just conspiracy theories being promoted by a known conspiracy theorist, journalist Nicky Hager. Hager had just released a book which provided evidence that the Prime Minister's Office was secretly using bloggers to attack the Opposition and their policies. What was interesting was the public's response to the Prime Minister's claim, "That's just a conspiracy theory." The public agreed that Hager's book was, indeed, a conspiracy theory, but then asked whether Hager's allegations were true. That is, just because it was a conspiracy theory didn't mean it was something the public ought not to take seriously.

Now, if the politics of a small Western democracy in the Pacific doesn't interest you, how about the lead-up to the invasion of Iraq in 2003 by the UK and the US? Both Prime Minister Tony Blair and President George W. Bush claimed critics of the invasion of Iraq were peddling "outrageous conspiracy theories." Why? Because these critics were suspicious that the intelligence dossier which provided the reason for the invasion had been doctored. According to the intelligence provided by MI5 and the CIA the Hussein regime in Iraq was manufacturing Weapons of Mass Destruction (WMDs). However, this intelligence went against everything the public had been told by United Nation's Weapon Inspectors, who were on the ground in Iraq at the time.

It turns out the so-called "conspiracy theorists" were right: the dossier was dodgy. It had been crafted in part to provide justification for what now looks like a pre-determined political decision by the US and the UK to invade Iraq and remove a regime said governments were uncomfortable with. Blair and Bush had used the label "conspiracy theory" in an attempt to shut down debate about the need to invade a foreign country.

These examples are but the tip of an iceberg. There are many more examples—like the conspiracy theory about the Moscow Trials of the 1930s, and the Watergate Scandal of the 1970s—which nonetheless turned out to be true. For plenty of examples of conspiracy theories which turned out to be true, see *Real Enemies* edited by Kathryn Olmsted.

The point here is, it's not clear that the common usage of "conspiracy theory" necessarily refers to something which is inherently suspicious. Indeed, it might even be an example of a kind of academic wisdom which is really just a scholarly superstition: perhaps we've been quick to dismiss conspiracy theories simply because we keep being told they are dangerous. But you don't have to just take my word for it: other scholars have done the hard work to analyze this assumption, and the results have been very interesting.

Psychologists in particular (like Michael J. Wood, and Lukić, Žeželj, and Stanković) have run surveys to find out how people react to something being labeled a "conspiracy theory," and it seems the ordinary public may not be entirely on board with thinking conspiracy theories are always irrational. This point has been made by other scholars as well, including sociologists, (like Orr, and Husting), anthropologists (Pelkmans and Machold), cultural theorists (like Bjerg and Presskorn-Thygesen), and historians (like McKenzie-McHarg and Fredheim).

So, if there is nothing inherently wrong with believing in conspiracy theories, then what is the problem with being a conspiracy theorist? But, then again, what are we to make of the notable conspiracy theorists like David Icke and Alex Jones, who seem to believe just about any old conspiracy theory they hear? Maybe it's not all conspiracy theories which are the problem. Rather, maybe it's just some conspiracy theorists!

You Don't Have to Be Paranoid to Believe Conspiracy Theories

One of the things we like to associate with conspiracy theorists is paranoia. Indeed, the most famous essay on this is Richard Hofstadter's "The Paranoid Style in American Politics." In that essay Hofstadter talked about conspiracy theorists as people who suffer from something which looked like paranoia, which he dubbed the "paranoid style." However, this was never intended to be a clinical diagnosis. Rather, Hofstadter was arguing that if we think paranoia is irrational, then something which resembles it—like belief in conspiracy theories—should be irrational as well.

Hofstadter accepted that conspiracies occur, so it was not by definition irrational to suspect the existence of conspiracies now. No, the problem was that conspiracy theorists saw conspiracies where none existed. The modern way of talking about the paranoid style is to talk about conspiracy theorists being "conspiracists" who suffer from the ailment of "conspiracism."

Conspiracism is the situation in which someone believes a conspiracy theory without reason. A lot of the academic discussion about conspiracism labels conspiracy theorists like David Icke and Alex Jones as "conspiracists."

Now, a conspiracist will obviously always be a conspiracy theorist. After all, to be a conspiracist requires that you believe a conspiracy theory. But not all conspiracy theorists will turn out to be conspiracists. This is because some conspiracy theorists will believe certain conspiracy theories for the right reasons. That is, they will be able to show that there really is (or was) a conspiracy behind some event.

This is not to say that conspiracism is not a potential problem: we can point towards people—remarkably similar to Jerry Fletcher or Alex Jones—who seem to believe any old conspiracy theory. But the question is whether these people (fictional or real) are typical, or are they just notable? After all, while we can point at people like Alex Jones or David Icke and say "See!" there are also a host of proponents of things which were called "conspiracy theories" by people who were called "conspiracy theorists" who were—nonetheless—right.

A Problem of History

One reason why we should embrace the idea we're all conspiracy theorists is that it means we can avoid a particular and tricky question—what are we to say about historical cases of actual conspiracy which were pejoratively labeled "conspiracy theories" at the time?

Here are two prominent examples from the twentieth century: the Moscow Trials of the 1930s, and the Watergate Scandal of the 1960s. The Moscow Trials saw the enemies of Joseph Stalin put on trial for being part of a conspiracy to bring his enemy, Leon Trotsky,. back to the USSR. Or, at least, those were the charges: it turns out the trials were a sham designed to legitimize a purge of Stalin's enemies.

The Watergate Scandal concerned not just a politically motivated break-in at the national headquarters of the Democratic Party in the US, but also the President and his advisors covering up their involvement in the affair.

Both of these are examples of conspiracies which happened at the highest level of government, and both are cases where the conspirators claimed conspiracy theorists were just peddling mere conspiracy theories. Now, the labeling at the time was likely just a ploy to cover up the conspiracy. That certainly fits with what happened. But the important thing to note is these conspiracy theorists weren't just accidentally right. They were putting forward something which the actual conspirators pejoratively labeled a "conspiracy theory."

However, by embracing the claim that—if we are historically or politically literate—we're all conspiracy theorists, then we can get around the tricky question of how we talk about conspiracy theories which turned out to be true. That is, if we remove the sting from both "conspiracy theory" and "conspiracy theorist," we can improve the quality of political debate by focusing on whether or not particular conspiracy theories are true or false.

Indeed, we can say something even bolder, which is that if we're being honest, we're all conspiracy theorists of some kind. The philosopher Charles Pigden has put this best: either you accept the reports in the news or in the history books that conspiracies have occurred, or you think that these reports are covering up what is really happening. Either way, you're a conspiracy theorist: you either accept that some theories about conspiracies have turned out to be true or you think there is some massive conspiracy going on right now to hide the truth! As such, any historically or politically literate person should consider themselves a conspiracy theorist.

But Aren't We (Also) All Conspiracists?

If we're all conspiracy theorists, aren't we all likely to be conspiracists as well? After all, many of us will believe some conspiracy theory for reasons which might not survive scrutiny.

This should not be a problem, however: if I turn out to be a conspiracist about, say, the events of September 11th 2001 (suppose I believe it was an inside job because one of my grandparents told me they saw that on a video on YouTube!), but have good reason to believe conspiracy theories like the Watergate Scandal, the Gulf of Tonkin incident, or the story behind the Moscow Show Trials (because I have read all the relevant history books), then I am a conspiracy theorist who just happens to also be a conspiracist about one particular conspiracy theory.

This is a bullet we should bite. If we accept that we're all conspiracy theorists, then it's not a stretch to think that sometimes we are going to be conspiracists. That should not be a shock to anyone's system: Most of us believe things that—on reflection—we have little to no evidence for. Some of these beliefs we naively get from other people and some of our beliefs end up being things which were convenient to believe given the circumstances we found ourselves in, rather than based upon the evidence.

So, it shouldn't be surprising that if we are conspiracy theorists, we're also likely to be conspiracists. What might be interesting is that we could turn out to be conspiracists about conspiracy theories which turn out to be true. After all, how many of us can really explain the conspiracy and cover-up at the heart of the Watergate scandal? How many of us know what the Dodgy Dossier contained, and how it was doctored and used by politicians to justify the invasion of Iraq in 2003?

Sometimes—when pressed—it turns out that we believe things without quite being able to explain why we believe them. But this is normal, and it is not a problem which is unique to conspiracy theories. So, rather than treat conspiracy theories as a special case, we should judge them like we judge any theory: on the evidence!

Alias Icke and Jones

Even if we bite the bullet, and admit that we might all be conspiracists, what are we to make of the Alex Joneses and David Ickes of our world? Surely, they are unusual examples of people who believe conspiracy theories?

Conspiracism, remember, describes an irrational belief in conspiracy theories, one where the conspiracist believes some conspiracy theory without reason. But it might be unfair to call Jones or Icke conspiracists. They may well be extreme examples of people who believe conspiracy theories, but it is not clear that they believe just any old conspiracy theory.

Icke, for example, has made it quite clear that he has changed his mind about theories in the past (he once thought he was the reincarnation of the Messiah, for example), and back in 2012 he was very vocal about not believing the so-called

"Mayan Doomsday" conspiracy theories which said the world would end in December of that year. Why? No doubt partly because those particular conspiracy theories contradicted the theories he did hold to be true.

While Icke's views might be unorthodox, he does have a systematic way of working out which conspiracy theories he believes to be plausible and those he does not. This is not a defense of Icke and his over-arching theory that we are controlled by humanoid lizards. Rather, the problem with calling Icke a conspiracist is that while Icke believes theories most of us think are implausible, he doesn't come to belief in those theories in an arbitrary or haphazard manner. Rather, what he thinks is evidence for a conspiracy simply differs from the rest of us.

The point here is that it's not clear that people like Alex Jones and David Icke are paranoid wrecks in the way that fictional portrayals like Jerry Fletcher and Edward "Brill" Lyle are made out to be. Unlike their fictional counterparts, Jones and Icke provide reasons and arguments for their chosen conspiracy theories. We might very well disagree with those reasons, but it is not at all clear that they are the conspiracists in the way certain scholars have claimed.

None Should Call It Conspiracism

A problem with critiques of belief in conspiracy theories as essentially "conspiracist" is they make out that belief in conspiracy theories generally is a problem by maintaining that any belief in a conspiracy theory is evidence of conspiracism.

It's hard not to think that such critiques end up assuming all conspiracy theories are questionable, or even false, without bothering to look at evidence for or against individual conspiracy theories. This is a problem because sometimes (and the jury is out as to how often this is the case) conspiracy theories turn out to be plausible and the kind of thing people should believe. From the Moscow Trials of the 1930s to the Dodgy Dossier in 2003, things which people called "conspiracy theories" ended up being the best explanation.

But conspiracist critiques rest upon the assumption that belief in conspiracy theories is suspect. Indeed, conspiracism almost looks like a view designed to explain why certain conspiracy theorists—those people we disagree with—are wrong. But if the kind of people at the heart of conspiracist critiques of belief in conspiracy theory turn out to be Jerry Fletchers and Edward "Brill" Lyles and not Alex Joneses or David Ickes, then we should avoid talk of conspiracism and focus on the arguments and evidence for or against specific conspiracy theories.

We do not want to be tilting against an illusory foe designed to make conspiracy theorists generally look bad. After all, if labeling something a "conspiracy theory" is enough to shut down debate about the possible existence of a conspiracy, then we make it all the easier for conspiracies to multiply and for conspirators to prosper. Given we know conspiracies do occur, and some conspiracy theories have turned out to be right, that is the kind of problem we ought to be worried about.

2

The Greatest Conspiracy Theory Movies

MARK HUSTON

> Although I'm certain that this will do nothing to discourage the con-
> spiracy peddlers: there is no evidence of a conspiracy.
>
> —Hammond Commission Spokesperson, *The Parallax View*

Oliver Stone's *JFK* (1991) is often considered the high-water
mark of conspiracy theory filmmaking. Movies following in its
wake include Ron Howard's *The Da Vinci Code* (2006) and the
uninspired, and unsubtly titled Mel Gibson vehicle, *Conspiracy
Theory* (1997). While *JFK* is an excellent movie, for my conspiracy theory money it is impossible to beat the one-two punch of
Alan J. Pakula's *The Parallax View* (1974) and *All the
President's Men* (1976). Not only are they both masterfully
made movies in their own right, but, when combined with modern philosophical and academic work on conspiracy theories,
they continue to provide insights well beyond the decade in
which they were made.

Of the two movies, *All the President's Men* is much more
well-known. Even cinephile friends of mine often have not seen
The Parallax View. However, via the helmsmanship of Pakula,
the two movies share a conspiratorial kinship that both encap-
sulates and moves beyond the decade in which they were made.
While *All the President's Men* is a story of the unfolding of a
historical event widely accepted as true, Watergate, and *The
Parallax View* is fictional, they both put the viewer into a similar set of emotional states: unease, paranoia, fear.

All the President's Men tells the story of the Washington
Post's Woodward and Bernstein, played by Robert Redford and
Dustin Hoffman in their acting prime. The movie follows the
reporters on a trail that starts with the Watergate break-in,

continues with meetings in creepy garage parking structures with the mysterious informant Deep Throat, and ending just before the final avalanche of stories that result in Nixon's resignation. Given that everyone knows the ending of the story before it starts and that real reporting is often slow and plodding, the movie probably should have been incredibly boring. Instead, it is absolutely riveting.

The Parallax View is riveting as well. *The Parallax View* is a fictionalized version of both Kennedy assassinations, with a senator being assassinated at the top of Seattle's Space Needle. It even includes a commission, clearly modeled on the Warren Commission with its "lone gunman" explanation, that bookends the film. The movie follows a relatively small-time newspaper reporter named Joe Frady, played by Warren Beatty in one of his best performances. Frady is there on the day of the assassination but does not witness it first-hand. However, an aide to the senator and Frady's former girlfriend does bear witness.

Three years after the assassination she visits Frady with worries that witnesses to the assassination are themselves being murdered. Soon thereafter she turns up dead, which launches Frady on a quest to expose the conspiracy. While investigating the death of one of the other witnesses, Frady finds a reference to the Parallax Corporation which, he figures out a bit later, manufactures political assassinations. Frady attempts to infiltrate Parallax but is instead ultimately set up as an assassin of another senator. The final scene of the movie shows a committee dubbing Frady yet another lone gunman.

What Is a Conspiracy Theory?

Conceptual analysis is one of the main jobs of philosophers. In order to be even remotely fair when discussing conspiracy theories, we need to try and provide a conceptual analysis of the term that balances the way "conspiracy theory" is used in the language without falling into the trap of automatically implying that all conspiracy theories are false. The great twentieth-century philosopher Ludwig Wittgenstein developed an ordinary-language approach to conceptual analysis that works very well with terms such as "conspiracy theory."

Wittgenstein rejects the notion that you can identify some kind of "essential" definition such as you might find in mathematics, and instead points toward overlapping criteria and paradigm examples to ground our conceptual understanding. Applying his method to the idea of conspiracy theories allows us to appreciate that conspiracy theories are on a sliding scale from completely bizarre, for example that shape-shifting lizard people control the world, to the very plausible and true: JFK and Watergate.

Some of the overlapping criteria for conspiracy theories that can be found in the academic literature include: they usually explain historical events, they run counter to the "received or official" view from the relevant authorities, they are typically driven by the intentions of a smallish group of conspirators, the intentions are for evil purposes, the conspirators operate in secret—that's why there are so many secret societies like the Illuminati—and they provide a narrative that can explain events that otherwise seem disconnected or irrelevant.

The granddaddy generator of modern conspiracy theories, and one believed by a very high percentage of our society, is the JFK assassination. The conspiracy theories attempt to explain JFK's assassination in terms of the malevolent activity of a group of conspirators, such as the mob, the CIA, Cuban exiles, while rejecting the official report from the Warren Commission. Unlike the crazy alien reptile conspiracies, the JFK conspiracy theories have unearthed a tremendous amount of information that many people find persuasive. In that sense, the best of the JFK conspiracy theories seem very close to the investigative journalism that has established everything from CIA assassinations to MK-Ultra programs; or to Senate hearings, like the Church Senate hearings, which established that many American reporters had also worked undercover for the CIA.

Event, Systemic, and Superconspiracies

The political philosopher Michael Barkun in his book *A Culture of Conspiracy* suggests that conspiracy theories fall into one of three categories: event conspiracies, systemic conspiracies and superconspiracies. These categories work very well when analyzing conspiracy theory films.

The JFK assassination film *Interview with the Assassin* (2002) is a perfect case of an event conspiracy. It follows a man who claims to have assassinated JFK from the grassy knoll and keeps us guessing whether or not he is a kook or actual assassin. Another JFK movie, the cult classic and very dark comedy, *Winter Kills* (1979), is a fantastic event conspiracy film. Like *The Parallax View*, it is clearly a fictionalized JFK assassination film. While it does bring in a range of conspirators, it remains focused on a few key figures and does not venture into the wide-ranging groups that you get in a movie like *JFK*. If you like conspiracy films, I highly recommend it!

The Da Vinci Code and National Treasure (2004) are good examples of superconspiracies. They both show long-standing historical conspiracies that span decades and the globe. They are clever and entertaining movies but not particularly interesting either filmically or conspiratorially. The movie *JFK*, a much better movie than the two just mentioned, may fit here as well. Though it seems like an obvious event movie, there are so many overlapping conspirators in the film that it presents a conspiracy that is better described as a super or systemic conspiracy.

Systemic conspiracies typically involve groups that engage in a series of event conspiracies in order to achieve some malevolent end, usually involving power and governments. Appeals to groups like the Freemasons or Knights of Malta, the Birchers' claims about Communists, and claims about the Illuminati would typically be categorized as systemic. The movie *JFK* may fit here as well. *JFK* is an interesting film because it really walks the line between being a systemic and being a superconspiracy theory. I think the further it leans toward being a superconspiracy the more implausible, and so less interesting, it becomes. A way of thinking of a superconspiracy is as an everything-but-the-kitchen-sink conspiracy. Though I love the movie, I think its greatness is dulled because it veers too far into the superconspiracy realm. One of the reasons I prefer *All the President's Men* and *The Parallax View* is because they are both perfectly contained which makes them even more effective than *JFK*.

Pakula's Paranoia

Identifying conspiracy theorists as engaging in a paranoid outlook on the world became the default analysis as a result of the historian Richard Hofstadter's influential essay, "The Paranoid Style in American Politics." That term often gets used when referencing Pakula's three masterpieces—*Klute* (1971), *The Parallax View*, and *All the President's Men*—dubbed the "paranoia trilogy." I won't talk about *Klute* here, since it doesn't directly relate to conspiracy theories.

Paranoia is a term often used to describe the atmosphere of the late 1960s and early 1970s. Encapsulated in the famous "For What It's Worth" line from Buffalo Springfield's song: "paranoia strikes deep; into your life it will creep." Pakula filmically demonstrates the general paranoiac mood captured in that song line. But, as the Joseph Heller chestnut expresses: "Just because you're paranoid doesn't mean they aren't after you." This notion is also captured by Pakula since, in both of the movies we're talking about, the conspiracies turn out to be real and the paranoid main characters are right!

One of the main reasons Pakula is so successful with these two movies is because they present conspiracies at the systemic level. In contrast, the event level is best for conjuring fundamental and basic emotions such as fear and sadness. Think of the Paul Greengrass 9/11 movie titled *United 93* (2006). That movie neatly captures the straightforward fear and terror of that day.

If the event level is best for evoking basic and intimate emotions such as fear, at the far other end we have the superconspiracy level. Superconspiracies work well for thrillers such as *The Da Vinci Code* or *National Treasure*. While on occasion these movies may momentarily engage deeper emotions, their sheer implausibility makes them ill-suited for any serious emotional depth. Superconspiracy films are akin to spy thrillers such as the James Bond or *Mission Impossible* movies. They're designed primarily for thrills and over-the-top fun.

The systemic conspiracy theory level is the one best-suited for evoking complicated, higher-order emotions such as paranoia, dread, and a general sense of dislocation. Another way of putting it is that movies at this level, when made well, create a mood or atmosphere. Again, by way of contrast, *JFK* walks the line between the systemic and supeconspiracy level. As a result, even though the film is quite excellent, it in no way creates the kind of mood and atmosphere that Pakula's two movies create.

I believe Pakula's two films are the best of this sort due to the combination of story, acting, and general filmmaking—such as the cinematography, the soundtrack, and so on. Other very good conspiracy movies that fall into this territory include Sydney Pollack's *Three Days of the Condor* (1975) and the only movie, in my view, that is at the level of Pakula's, John Frankenheimer's *The Manchurian Candidate* (1962). Other than that movie, I take it that Pakula's films are by far two of the greatest.

Conspiracy Film Genre

As a genre, we need to separate out conspiracy films from other types of films that may be similar. Many movies portray conspiracies of some sort or other. This element is the basis of most crime films, mysteries, and thrillers. However, for a work to rightfully be categorized in the genre of conspiracy theory film, the film has to give us a systemic conspiracy or superconspiracy where the conspiracy satisfies most of the overlapping criteria mentioned earlier—the film must pose an explanation that runs counter to the mainstream knowledge-conferring author-ities, the conspiracy must explains some events based upon the evil intentions of a relatively small group, and so forth—and the conspiracy in the movie must be one typically understood to be a conspiracy by the culture at large. That is why *JFK* or *Three Days of the Condor*, which suggest a CIA conspiracy, are perfect examples that fit the genre.

The Parallax View also fits the paradigm of a conspiracy theory movie by presenting a global corporation as the driver of political assassinations. *All the President's Men*, though, is slightly less obvious. Yet, it too, is properly understood as a con-spiracy theory film. Given that we know that Watergate is true, some might be reluctant to call Watergate a conspiracy theory. But, given that Watergate meets virtually all of the overlap-ping criteria for a conspiracy theory, the only possible reason to reject it would be as a result of a question-begging definition. On any reasonably fair definition of 'conspiracy theory' Watergate counts, which is why most academic works on con-spiracy theories refer to Watergate as a true, or "warranted," conspiracy theory.

All the President's Men

A main element that conspiracy theory, con-game, and heist films often have in common is keeping the audience off balance. This is especially salient in the better con films, such as those made by David Mamet (for example, *House of Games* and *The Spanish Prisoner*). One of the primary differences, though, is that in con and heist movies there is usually a protagonist who ultimately prevails because she's able to think several steps ahead of the antagonists in the story. Usually that is not the case in conspiracy theory films, which is why they often leave the viewer unsettled when the movie is over.

All the President's Men seems to run contrary to the point I just made since we know all along that Woodward and Bernstein ultimately prevail. However, and this is a point rarely made about the movie since most analyses focus on the successful journalism presented, the aesthetics of the movie suggest a long-term dread that is not quenched merely because the president resigns.

When viewing *All the President's Men*, we get the sense of the main characters being watched or observed, that there is a presence, if you will. That technique works to establish the overall mood of paranoia and dislocation. Almost all of the shots of the White House are at night and from a distance, suggesting that the Capital is shrouded in darkness and secrecy. When Woodward meets Deep Throat in the parking garage, again at night, you watch Woodward from a distance walking down the stairs to get to the garage with the only lighting coming from the stairwell lights. The perspective taken is as if you are observing him from the vantage point of, perhaps, someone watching him within the movie. Because Pakula constantly leaves shots like this ambiguous—is someone else observing him or not?—the viewer's sense of dread keeps increasing as the stakes get higher.

Additionally, Pakula uses a lot of still-camera shots with deep focus so the viewer can see everything in the frame whether far away or not, and in those shots he allows important action to take place in off-center locations or even on small televisions that are in the news office. Since most movies highlight and focus dead center on important elements, shots of the sort Pakula uses keep the viewer off-balance which merely adds to the mood of oddness and dislocation where the world is not quite how it normally seems.

One of Pakula's most famous shots does several things in the movie. At one point Woodward and Bernstein are looking over documents in the Library of Congress. Initially, we see a close-up from above of the two of them working furiously. Then the camera pulls up and away, apparently from the middle of the Library. Each time the camera pulls away we see more of the circular building, more people, but eventually we are unable to recognize which of those people are Woodward and Bernstein. Apart from the fact that the shot is a virtuoso performance, it shows how small the workings of just two people can be—like salmon swimming upstream. It also gives the sense of futility because, apart from the immensity of the building, the circular shape leaves one with the feeling that they, and we the viewers, are on a giant hamster wheel ultimately going nowhere. Another important mood is that of lack of control for the "little guy," but complete control for the conspirators, that is present in this shot and throughout the film. As the famous conspiracy theorist and *Illuminatus Trilogy* author Robert Anton Wilson often ominously said, "Everything is under control."

The Parallax View

The Parallax View is even more disconcerting than *All the President's Men*. Again, Pakula uses long shots where important events happen yet we only view them from a distance and we are not privy to all information. Plus, witnesses keep dying, reminiscent of the way many people related to the Kennedy assassination died. The improbability of this is often emphasized by JFK conspiracy theorists. Finally, and unlike most movies and *All the President's Men*, not only does the main character not succeed in taking down the bad guys, but he's actually set up as a fall guy and murdered at the end.

Along the way, Frady finds layer after layer of people involved in the conspiracy. He also realizes that the corporation involved in the assassinations is global and has been engaging in this activity for a while; thus, we see the systemic nature of the conspiracy. Even when Frady does get help, for example from his newspaper editor, the people who help him end up dead.

All of these elements are more than successful in creating an atmosphere of dread and paranoia. However, it is when Frady infiltrates the Parallax Corporation where we get a truly great film moment. Frady pretends to be a sort of sociopath and he figures out a way to get recruited by Parallax. He goes to their offices and is escorted into a room with just a big chair and a big screen, a bit evocative of the brainwashing scene in *A Clockwork Orange* (1971).

Once Frady sits down, music starts to play and then images start to appear on the screen. Initially the music is pleasant as are the scenes (moms, dads, children, parades and so on). The images are interspersed with text in big letters: LOVE, MOTHER, FATHER, ME, HOME, COUNTRY, GOD, ENEMY, and HAPPINESS. The images and text change more and more rapidly, with the music becoming discordant, which constantly changes the context of how to understand the terms being used. For example, a father holding his baby followed by the text LOVE, then a father chasing a young child with a belt followed by FATHER. This is purportedly used to test whether Frady should be used as an assassin. But you come to figure out later that Parallax was manipulating Frady the entire time. The movie critic Matt Zoller Seitz considers this to be one of the greatest bits of film editing in American movies and I agree.

The movie builds toward a typical Hollywood ending where the bad guys are revealed and the good guy wins. Instead, though, it pulls the rug out from under you and leaves the conspiracy in place while having the protagonist die. This is akin to classic Aristotelean tragedy where viewers are left feeling pity and fear. However, with *The Parallax View* the viewer is left feeling fear, paranoia, and dread—especially since this seems to reflect the way the world, and a conspiracy of this sort, often works, as illustrated in many historical examples such as MK-Ultra or CIA coups in other countries.

The Conspiracist Mentality

The philosopher Quassim Cassam has identified a character trait that he calls the *conspiracist mentality*. Cassam considers the conspiracist mentality to be a vice linked to traits such as prejudice, gullibility, and cynicism. For our purposes, though, it is often the case that in conspiracy movies the conspiracist mentality usually pays off, at least in terms of revealing an actual conspiracy.

Frady clearly exhibits the conspiracist mentality including being admonished by his editor early in the film for a previous story he pursued. Plus, that mentality is turned against him by the report at the end of the movie where Frady is said to have been "obsessed" with the assassination of the senator which lends credibility to his becoming an assassin himself.

Comparing Frady to Woodward and especially Bernstein is illuminating. Like Frady, Bernstein often makes leaps of logic where he assumes a conspiracy without having evidentially establishing one. That is partly why he and Woodward work well together. Bernstein makes inferential leaps that are often correct, but Woodward tethers him to proper evidential sources as do the editors of the paper. And while all of the characters are somewhat paranoid, where Woodward and Bernstein do the plodding work necessary to do good reporting, Frady goes in disguise and acts like sociopath in order to try and get at the truth.

This is one of the major differences between most real-world conspiracy theorists and investigative journalists who are attempting to uncover a conspiracy. Solid epistemic work, whatever the field, must be grounded in proper sources of evidence and in, as Cassam puts it, proper modes of inquiry. Good journalists trying to reveal a conspiracy present a preponderance of sources and evidence. The serious conspiracy theorists do the same, they do not merely look up a bunch of different sources online and make incredible leaps of logic.

The best conspiracy films work at the systemic level with a main character who, while exhibiting a conspiracist mentality, ultimately grounds their decisions in good evidence. Again, by way of contrast, the leaps made by Jim Garrison, played by Kevin Costner in *JFK*, are often so ill-grounded that they make him seem as if he is merely trying to create a good story rather than reveal the truth. That is the opposite of Woodward and Bernstein, and even Frady.

That's just one of many reasons why I consider Alan J. Pakula to be the greatest conspiracy theory movie director.

3
Everyone's a Conspiracy Theorist

CHARLES PIGDEN

According to the conventional wisdom, to label someone a *conspiracy theorist* is to imply that they are some kind of kook and to label a theory as a *conspiracy theory* is to suggest that it is a theory that it isn't rational or respectable to believe or maybe even to investigate.

Indeed, to call somebody a "conspiracy theorist" or their ideas "conspiracy theories" is (usually) to make a *silencing* move. To describe somebody as a conspiracy theorist is to suggest that their deluded burblings can be safely dismissed; to describe a theory as a conspiracy theory is usually to suggest that it is way too silly to be worth considering, that it should never have been mooted in the first place, and that it can safely be banished from the domain of rational debate.

Conspiracy theories, simply because they *are* conspiracy theories are intellectually beyond the pale, and the same goes for the people who believe them. Thus, the conspiracy theorists should shut up and any attempts to discuss their theories should be shut down. That's why mainstream journalists, timidly inclined to suggest a conspiracy theory, often preface their remarks with "I'm not a conspiracy theorist but . . ." They're uneasily aware that the theory they are about to suggest involves a conspiracy (and is therefore, presumably a conspiracy theory), but they want to resist the implication that they have thereby taken leave of their senses. The conventional wisdom is that it is deeply unwise to be a conspiracy theorist which is why the conventionally wise are so keen to distance themselves from the accusation.

Why Conspiracy Theories as Such *Have* to Be Suspect

But is the conventional wisdom correct? Are conspiracy theories *as such* somehow suspect or unbelievable? *Some* are, of
course since some wear their absurdity on their face, but that
is not the point at issue. The claim is not that *some* conspiracy
theories are silly or even that *many* of them are silly, because
that is something nearly everyone concedes.

The claim has to be that conspiracy theories are *always* or
almost always suspect or unbelievable and that this is not
because of special features of this or that conspiracy theory
(that the conspirators did not know each other; that they could
not have met as they were alleged to have done; that the
alleged conspiracy would not have been in the conspirators'
interests; that the theory endows the alleged conspirators with
magic powers), but because there is something *in the nature of
conspiracy theories* that makes them almost always not only
false but ridiculous. For unless something like this is true the
slur simply does not make sense.

Suppose I suggest (on the basis of the Mueller Report) that
Trump conspired, rather incompetently, to obstruct the course
of justice. And suppose somebody responds "But that's a conspiracy theory!" Unless conspiracy theories *as such* are somehow suspect or unbelievable (and are widely known to be so), I
can simply reply "So what? Sure, it's a conspiracy theory, but
people often conspire and there is plenty of evidence in the
Mueller Report that Trump did just that! Yadda, yadda,
yadda..." It is *only* if conspiracies are almost always false or
unbelievable (and are widely known to be so) that this sort of
response is ruled out. The same goes for "conspiracy theorist"
when it is used (as it is almost always used) as a slur.

Roger Stone's rather unprepossessing sidekick, Jerome
Corsi, is regularly described as "a right-wing conspiracy theorist" as if that in itself were enough to damn him. But unless
conspiracy theories *as such* are suspect and unbelievable it
obviously is not. The implication, of course, is that Corsi's conspiracy theories are false and fantastic and maybe even dangerous and that he is therefore a disreputable person for
peddling them in the first place. But that only follows from the
claim that Corsi is a conspiracy theorist on the (unspoken)
assumption that conspiracy theories in general are false and
fantastic. If that assumption is false (or is not generally con-
ceded) then more needs to be said in order to establish that
Corsi is the political low-life and borderline loon that he is
widely alleged to be. Instead of "Jerome Corsi, Stone ally and

right-wing conspiracy theorist" (which is how he is described on my first Google hit) you would have to say something like "Jerome Corsi, Stone ally and purveyor of false and fantastic right-wing conspiracy theories" (which is a little less snappy but a lot more alliterative).

The fact that journalists in general don't think that they have to say this sort of thing (preferring the snappier phrase "conspiracy theorist") shows that many of them subscribe to the myth that conspiracy theories are almost always suspect, false, and unbelievable, simply because they *are* conspiracy theories.

Not Necessarily Suspect

But it's a myth that conspiracy theories are almost always suspect, false, and unbelievable. There's nothing intellectually wrong with conspiracy theories as such. Some of them (maybe even many of them) are false and fantastic, some of them (as we shall see) are actually dangerous (that is, they suggest the kind of ideas that can kill); but they are not false, fantastic, or dangerous *because* they are conspiracy theories, but because they are false, fantastic, or dangerous. It is simply *not true* that conspiracy theories, simply by being conspiracy theories, are suspect, false, or unbelievable. Indeed this is not just false: it is (as we shall see) a *dangerous* falsehood, a myth that can be, and has been, weaponized to catastrophic effect. Let's start with the falsehood first.

Conspiracy theories are theories which postulate or suggest conspiracies. People frequently conspire. Thus, many conspiracy theories are true. Furthermore, *when* people conspire there is often good reason to think that they have done so. Thus, many conspiracy theories are rationally believable. Hence many conspiracy theories are both true and believable. It is therefore false to suppose, that conspiracy theories, simply by virtue of *being* conspiracy theories are suspect, false, or unbelievable—which means that the conventional wisdom is deeply mistaken.

Furthermore, a conspiracy theorist is simply a person who subscribes to some conspiracy theory. Since many conspiracy theories are not only true but the kind of thing that it is rational to believe, there need be nothing irrational about being a conspiracy theorist (though of course, many conspiracy theorists are indeed irrational). Indeed, since history and the nightly news are full of well-authenticated conspiracy theories, and since politically literate people are disposed to believe both history and the nightly news (though they may of course take them with a suitable pinch of salt), it follows that every politically literate person is a big-time conspiracy theorist. But if every politically literate person is a big-time conspiracy theorist, then it can't be irrational to be a conspiracy theorist.

Finally, "conspiracy thinking" is simply the thinking that leads people to postulate or adopt conspiracy theories. Since it's perfectly rational (indeed a condition of political literacy) to subscribe to at least *some* conspiracy theories, there need be nothing irrational about conspiracy thinking. Indeed, the thoughts which lead people to formulate and adopt conspiracy theories are so very diverse (some rational, some irrational, some desire-driven and some insane) that there isn't anything very sensible to be said about "conspiracy thinking" in general.

However it's not just wrong but *dangerous* to suppose that conspiracy theories *as such* are too silly be believed or investigated. For the fact is that this idea is deployed selectively and dishonestly, sometimes with catastrophic effects. There is a real-life case of an actual conspiracy in which the conspirators deployed anti-conspiracist rhetoric to protect their schemes from public scrutiny and criticism whilst pushing conspiracy theories of their own which were just as false and even more dangerous than conspiracy theories in general are alleged to be. In consequence of this conspiracy (not to mention the false conspiracy theories) hundreds of thousands of people died.

Tony Blair and the Iraq War

The terms "conspiracy theory" and "conspiracy theorist" are often used as silencers to shut down debate or to dismiss an opponent as not worth bothering with. Tony Blair, former Prime Minister of Great Britain was a champion exponent of this tactic even though he was both a big-time conspirator and a big-time conspiracy theorist. The Iraq War, as argued for by Bush, Blair, and their acolytes, was officially based on *three* conspiracy theories, the first true but the other two false.

1. That the events of 9/11 were due to a conspiracy on the part of al-Qaeda (who were themselves in league with the Taliban).

2. That the regime of Saddam Hussein was in cahoots with al-Qaeda, making him in some sense an accessory to the events of 9/11.

3. That the regime of Saddam Hussein had *successfully* conspired to evade the UN inspectors and had acquired (or

retained) weapons of mass destruction and perhaps was on the way (via the acquisition of yellowcake from Niger) to gaining a nuclear capability, thus making the regime a clear and present danger both to the UK and to the US.

These weren't *described* as conspiracy theories, of course (though they clearly postulate conspiracies), perhaps because the alleged conspirators were foreign folk from far, far away, and perhaps because the proponents of these theories were noble patriots who could not reasonably be suspected of peddling false conspiracy theories to further a war-mongering agenda. But as we now know they were in fact peddling false conspiracy theories to further a war-mongering agenda focused on regime change.

Furthermore, there was definitely a conspiracy to *talk up* the threat of Iraqi WMDs even if Bush and Blair managed to take themselves in with their own propaganda. (It's pretty clear now that they preferred to listen to intelligence sources that told them what they wanted to hear and refused to listen to what they did not want to be told. One is tempted to paraphrase Cole in the *The Sixth Sense*. "I see war-mongering people. Walking around like regular people. They only see what they wanna see. They only hear what they wanna hear. They don't even realize what warmongers they are." Of course, the consequence of Bush and Blair's selective inattention was that thousands of people wound up dead.) We know about the conspiracy to talk up the threat from government memos—for instance Powell's memo to Bush prior to the Crawford meeting and the subsequent Downing Street Memo:

> Blair continues to stand by you and the U.S. as we move forward on the war on terrorism and on Iraq. He will present to you the strategic, tactical and public affairs lines that he believes will strengthen global support for our common cause. . . . On Iraq, Blair will be with us should military operations be necessary. He is convinced on two points: the threat is real; and success against Saddam will yield more regional success. (Powell to Bush, 28th March 2002)

> Military action was now seen as inevitable. Bush wanted to remove Saddam, through military action, justified by the conjunction of terrorism and WMD. But the intelligence and facts were being fixed around the policy (Downing Street Memo, Rycroft to various ministers, 23rd July).

My point, however, is that both the *actual* conspiracy (to justify the war by talking up the threat of WMDs) and the false conspiracy *theories* were defended by invoking the myth that conspiracy theories as such are false and unbelievable. Blair began by using this tactic to try to get the Labour Caucus onside:

> Tony Blair today derided as "conspiracy theories" accusations that a war on Iraq would be in pursuit of oil, as he faced down growing discontent in parliament at a meeting of Labour backbenchers and at PMQs. ("Blair: Iraq Oil Claim Is "Conspiracy Theory'," *Guardian*, 15th December 2003)

Well, if it wasn't about oil it was certainly about regime change (which would have made the oil more readily available to US interests) and, as we now know, there was indeed a conspiracy to sell *that* to the public by talking up the threat of WMDs. Not only did Blair use the rhetoric of conspiracy theory denialism to defend the war against perfectly reasonable suspicions before he had even launched it, but he also used it afterwards to defend his own actions when everything had gone very visibly pear-shaped:

> Appearing on US Fox News's Huckabee show, Mr. Blair spoke for the first time publicly since he gave evidence to Sir John Chilcot last month. The former premier said: "There's always got to be a scandal as to why you hold your view. There's got to be some conspiracy behind it or some great deceit that's gone on. People just find it hard to understand that it's possible for people to hold different points of view and hold them reasonably." ("Tony Blair: Britain Is Obsessed with Iraq Conspiracy Theories," *Evening Standard*, 8th February 2010)

And of course there is the time that he shrugged off the suggestion (contained in a secret memo for which the leakers were being prosecuted) that he had had to dissuade his buddy George Bush from bombing the al-Jazeera headquarters in Doha because he did not appreciate their coverage of the Fallujah campaign: "Look, there's a limit to what I can say—it's all sub judice, . . . But honestly, I mean, conspiracy theories . . ." (*Daily Telegraph*, 28th November 2005).

Thus, the idea that conspiracy theorists are suspect was used by the proponent of two false conspiracy theories to deflect attention from his own conspiracy, thus helping to justify a war which resulted in hundreds of thousands of unnecessary deaths. Conspiracy theories can kill, but so can the idea that they are always or almost always, false.

What Exactly *Is* a Conspiracy Theory?

A conspiracy is a secret plan to influence events by partly secret means. Conspiracies do not have to succeed to be conspiracies nor do they have to remain secret to be conspiracies. Failed conspiracies are still conspiracies. The definition is neutral. It does not imply that conspiracy theories are necessarily defective, nor does it imply that they are not.

Stauffenberg's conspiracy to assassinate Hitler in 1944 still counts as a conspiracy despite its lack of success. And conspiracies that have been exposed or betrayed or which the conspirators were subsequently happy to acknowledge are still conspiracies.

The conspiracy to murder Julius Caesar (planned and partly executed in secret as the daggers had to be smuggled into the Senate House) did not cease to be a conspiracy just because once Caesar had been successfully assassinated the conspirators very publicly gloried in the deed. Though it was crucial to Napoleon's conspiratorial coup of the 18th Brumaire that it should remain largely secret until it had been enacted, there was no need to keep it secret once it had succeeded and the young General Bonaparte had become First Consul (in effect dictator) and subsequently Emperor of the French.

I will add one further condition to my definition of a conspiracy as *a secret plan to influence events by partly secret means*, that is, the cancellable implication that the plan in question is *morally suspect, at least to some people*. This is not to say that conspiracies are automatically evil—there can be noble or patriotic conspiracies such as Stauffenberg's plot to assassinate Hitler—but we do not normally call a secret plan a conspiracy if it is morally innocuous. For instance, we would not normally describe the executives of Mazda as "conspiring" to bring out a new model Miata, because although the product launch is likely to be planned and partly executed in secret (to avoid giving commercially sensitive information to competitors), nobody is likely to regard this as morally objectionable (unless, like the executives of Volkswagen, they were also plotting to evade environmental regulations). By contrast the Stauffenberg Plot would have seemed morally objectionable at least to some people, namely to dedicated Nazis.

So much for *conspiracies,* what about conspiracy *theories*? A theory is a set of propositions which purports to explain some alleged facts. As every scientifically literate person knows, a "theory" can be true or false, sensible or silly, speculative or well-confirmed, decisively refuted or proven beyond all reasonable doubt. The Flat-Earth Theory is a theory even though it is utterly idiotic and the theory that the Earth goes around the sun (rather than vice versa) is a theory even though it is a theory that no sane and educated person would care to dispute. Thus, a conspiracy *theory* is a theory which purports to explain some alleged facts by positing a conspiracy: a conspiracy being a secret plan to influence events by partly secret means in a way that some people are likely to regard as morally objectionable.

Since conspiracy theories are *theories* and since theories in general can be true or false, sensible or silly, speculative or well-confirmed, decisively refuted or proven beyond all reasonable doubt, the term "conspiracy theory" carries no implications, one way or the other, about the status of *actual* conspiracy theories. So far as the meaning of "conspiracy theory" is concerned, conspiracy theories can be true or false, sensible or silly, speculative or well-confirmed, decisively refuted or proven beyond all reasonable doubt.

Furthermore, since conspiracy theories are *conspiracy* theories, and since conspiracies are secret plans to influence events by secret means *which may or may not be successful* and *which may be no longer secret*, the term "*conspiracy* theory" carries no implication that the alleged conspiracies are either successful in themselves or successfully secret.

And so, a conspiracy theory is still a conspiracy theory if the conspiracy that it claims to exist is alleged to have been a dismal failure or if the conspiracy that it claims to exist has ceased to be secret. This is important because some British commentators claim to favor cock-up theories over conspiracy theories (where "cock-up" is a term in British English, corresponding more or less to "screw-up"). Conspiracies and cock-ups are not incompatible. If I'm not trying to do something, I can't cock it up, and if the thing I'm trying to do is execute a secret plan by partially secret means, then if I cock it up I will have cocked up a conspiracy.

Given my definition of "conspiracy theory" it is logically possible that conspiracy theories in general are false and fantastic, as would be the case if nobody ever conspired to do anything. But it is equally possible that many of them are both true and believable, as would be the case if people often conspired, and if there was often good evidence that they had actually done so. The definition by itself, says nothing either way. It is the nature of reality that determines whether conspiracy theories in general are silly and suspect or whether enough of them are true to dispose of the conventional wisdom.

Conspiracies Galore

Now that we have defined our terms, we can turn to the substantive issue: are conspiracy theories as such intellectually defective? And the answer is clearly "No." For the fact is that people frequently conspire and when they do there is often good reason to suppose that they have done so. It is difficult to mount a coup without conspiring, so almost every coup, successful or otherwise, is due to a conspiracy, the details of which often become widely known.

Thus, the theory that Napoleon conspired to overthrow the *Directoire* in 1799 is not open to serious doubt, nor is the theory that his putative nephew Napoleon III conspired to overthrow the Second Republic in December 1851. The same goes for the theory that the Bolsheviks conspired to overthrow the Provisional Government in 1917 or the theory that the collapse of Communism in the USSR was partly due to the failed coup by the "Gang of Eight" in 1991.

But it isn't just coups. Although there sometimes are lone gunmen, acts of terror and political assassination are typically conspiratorial affairs. So too is espionage which often has a major impact on world affairs. For instance, the USSR's rapid success in building an atomic bomb following Hiroshima, was much facilitated by atom spies such as Klaus Fuchs and Alan Nunn May, and this in turn led to an arms race which intensified the Cold War. Organized crime is essentially conspiratorial, and if it is sufficiently successful, it can have a major impact on the social scene. Cover-ups too are conspiratorial affairs. Even political "spills" in democratic parties are often pretty conspiratorial with a lot of furtive feeling out of potential supporters before the dissidents have the numbers to challenge the party leader. (In my own county, New Zealand, in 1997, the then Prime Minister Jim Bolger was dethroned by a party room coup organized by his successor, the Transport Minister, Jenny Shipley.) Other conspiracies are far more sinister.

"Disappearances" and other acts of repression are usually planned and partly executed in secret since if you want to kidnap the regime's opponents before consigning them to a shallow grave it is better not to let them know that you are coming. Stalin's Great Terror was a gigantic conspiracy to do away with potential opponents by fabricating a set of *false* conspiracy theories which were subsequently used to convict his victims. It should not be forgotten that the Holocaust was planned and executed in relative secrecy and was therefore a case of genocide by conspiracy.

It is sometimes suggested that although conspiracies were common once upon a time, and although they are still perhaps rife among benighted foreigners, they are relatively rare in modern democratic counties. This is certainly not true of the USA, which has had a long litany of well-authenticated conspiracies over the last seventy years. Here are just a few involving government agencies:

1. The COINTELPRO program on the part of the FBI to subvert "subversive" (but legitimately democratic) organizations by disinformation campaigns including anonymous letters to the husbands of white civil rights activists suggesting that their husbands were having affairs with their black comrades.

2. MKULTRA, a conspiratorial CIA program designed to investigate the possibilities of mind control involving prostitutes who lured unsuspecting johns back to their (carefully bugged) apartments where they would be administered LSD and other drugs.

3. The Congress for Cultural Freedom, a supposedly independent anti-totalitarian organization that was actually run by the CIA.

4. Watergate and all the many associated conspiracies on the part of both CREEP (Committee for the Re-Election of the President) and the Nixon administration, including the ensuing cover-ups.

5. The Iran-Contra affair, in which it emerged that people in the Reagan administration had secretly sold arms to Iran and tried to direct part of the proceeds to support the Contra Rebels in Nicaragua.

6. The "extraordinary rendition" scheme under Bush, whereby foreign nationals suspected of terrorism were kidnapped and flown to foreign places where they could be subject to enhanced interrogation techniques (torture) that were illegal in the US.

We can see that conspiracy theories are sufficiently common (and in many case sufficiently well-proven) to discredit the conventional wisdom that conspiracy theories *as such* are suspect and defective. But, in fact, we can go one better.

We're All Conspiracy Theorists Now

To be a conspiracy theorist is to subscribe to a conspiracy theory. Coups, acts of terror, assassinations and cover-ups, espionage and many kinds of political party intrigues, not to mention the activities of organized crime, are all typically conspiratorial enterprises. Both history and the nightly news are absolutely full of such things.

Thus, you cannot coherently believe that history and the nightly news are more or less correct without subscribing to a vast array of conspiracy theories. Every politically literate person—everyone with some knowledge of history who pays a modicum of attention to the news—is by virtue of their political literacy a big-time conspiracy theorist even though they are usually unaware of that fact.

Indeed, the only way to *avoid* being a conspiracy theorist is to render yourself politically illiterate, either by resolutely leaving the history books unopened and studiously ignoring the news or by adopting an attitude of equally resolute skepticism to every item of political or historical information that happens to come your way. But to adopt either of these strategies would be highly irrational, a sort of intellectual suicide. Now if every politically literate person is a big-time conspiracy theorist, subscribing to a wide range of conspiracy theories, then it can't be irrational to be a conspiracy theorist. Indeed, since it would be irrational *not* to be politically literate, a stronger conclusion follows.

Since it would be irrational *not* to be politically literate, and since you can't be politically literate without being a conspiracy theorist, then it isn't just that it is not irrational *to be* a conspiracy theorist (by subscribing to some well-authenticated conspiracy theories); it *is* irrational *not* to be one. The conventional wisdom could not be more wrong.

4
How to Build a Conspiracy Theory

BRETT COPPENGER AND JOSHUA HETER

Conspiracy theories are weird. On the one hand, they pique our interest and demand our attention; on the other hand, we realize that we should probably know better than to accept them (or to even pay them much attention) because of their inherent dubious and often unseemly nature. Conspiracy theories are like intellectual car wrecks; we know we *should* look away, but as we pass them on the epistemic highway, it can often be too difficult not to just gawk at them.

Why is this the case? What explains the tension here; why is it that we can't look away even though we know we should? The answer seems to be that conspiracy theories are so appealing because they do an amazing job of explaining (or connecting) seemingly unrelated bits of data. However, despite this—despite how easy conspiracy theories are to get off the ground because of this fact—it nevertheless still turns out that they ultimately should be rejected, at least in most typical cases.

For better or worse, in order to fully understand the intuitive appeal of conspiracy theories and why they really should be ultimately disregarded, we'll have to do a bit of mathematics. But, don't worry, the math is surprisingly easy. And, as a bonus, if you master it you will sound intelligent in front of your friends.

So, You're Saying There's a Chance

Probability theory is the very practical study of scenarios in which we are trying to determine the most likely outcome in a larger set of possible outcomes. Probability theory isn't just useful to gamblers, it's something anyone can employ in their everyday life. And, as it turns out, some very simple principles in arithmetic can teach us a fair amount about philosophy.

In order to understand the intuitive appeal of conspiracy theories, we'll first need to consider a very useful theorem for dealing with *conditional probabilities*: Bayes's Theorem. Believe it or not, you are probably (excuse the pun) very familiar with conditional probabilities, you just don't know it yet. Consider an example: if I were to ask, "How likely is getting heads if I flip a coin?" You would, of course, jump at the answer: fifty percent; the *probability* of getting heads is fifty percent, or .5 on a scale from 0 to 1.

Calculating the probability of the outcomes of coin flips should seem relatively straightforward, but it's worth pointing out that the probability of a coin flip being heads being .5 is really a *conditional* probability. It depends on the *condition* that the coin is fair (that it is not a trick coin). So, really, the probability of heads *given* a fair coin is .5. And, of course, the probability of tails given the same coin is also fifty percent or .5 (on that same scale from 0 to 1).

Most conditional probabilities are not as easy to deal with as are fair coins. But, Bayes's Theorem—our theorem for dealing with conditional probabilities—makes the job much, much easier. Bayes's Theorem is a theorem that allows us to determine how relevant some piece of evidence is for a hypothesis. It demonstrates just how likely a hypothesis is to be true in light of a certain bit evidence.

The dirty part: take 'P' to mean probability, 'h' to be a variable for some hypothesis, and 'e' be a variable for some evidence. According to Bayes's Theorem the probability of some hypothesis given some new evidence (P(h|e) = the probability of h given e, also known as the Posterior Probability) is equal to the prior probability of that hypothesis (P(h)) multiplied by the likelihood of getting that evidence given the hypothesis (P(e|h)), divided by the expectedness of the evidence (i.e., P(e)):

$$P(h|e) = \frac{P(h) \times P(e|h)}{P(e)}$$

or, in English $$Posterior = \frac{Prior \times Likelihood}{Expectedness}$$

Easy right?

Here's the point, what Bayes's Theorem makes explicit is that when it comes to evaluating the reasonableness (or the probability) of some hypothesis, we need to pay attention to two different things.

First, we have to pay attention to how probable the hypothesis is on its own (the prior probability) before the evidence ("e") is introduced.

Second, we must pay attention to how much learning of e would raise the probability of h, which is dependent on two additional factors: how strongly the hypothesis predicts the evidence (the likelihood: $P(e \mid h)$) and how expected e is (the $P(e)$).

Thus, according to Bayes's Theorem, when it comes to evaluating how likely a hypothesis is given (or in light of) some evidence (the posterior probability) we really need to be sensitive to two things: the prior probability of the hypothesis within the realm of feasibility (is it at all likely?) and what can we say about the likelihood ratio (likelihood / expectedness, or $P(e \mid h)/P(e)$)?

Why You Can't Look Away

Now that the machinery is in place, we are in a position to understand the intuitive appeal of conspiracy theories. Here is our diagnosis. Conspiracy theories exploit what is called the base rate fallacy. The reason why conspiracy theories are so appealing is because they do an amazing job of explaining (or connecting) seemingly unrelated bits of data. In other words, when it comes to a conspiracy theory, the $P(e \mid h)$ is ridiculously high.

If some hypothesis is true (the conspiracy theory), we would have the perfect explanation of some bit of evidence. So, in other words, the likelihood ratio is exceptionally top heavy. Good conspiracy theories typically entail the evidence in question. The problem, though, is that the prior probability of the hypothesis is so low that it swamps out the significance of the likelihood ratio.

Again, conspiracy theories usually do a really good job of making seemingly unrelated and random events seem very predictable and tied together. They make observations/evidence that would normally be very unlikely seem very likely. Or, in other words, the likelihood ratio is very top heavy:

$$\frac{P(e \mid h)}{P(e)} \quad = \quad \frac{[\text{really high}]}{[\text{low}]}$$

The problem with conspiracy theories is that the prior probability is so low, we can screen it off as unimportant:

$$P(h) = [\text{super-super-duper low}]$$

But that means we are left with the following calculation:

$$P(h|e) = \frac{[\text{super-super-duper low}] \times [\text{really high}]}{[\text{low}]}$$

And if the prior probability of a theory is small enough (super-super-duper low), its prediction of unlikely/unusual events is not that spectacular (really high/low is not significant).

Some Well-Known Conspiracy Theories

Conspiracy theories typically get off the ground because they do an amazing job of explaining (or connecting) seemingly unrelated bits of data. Consider the following cases which seem to do just that.

Michael Jordan's Suspension

The theory: Michael Jordan's first retirement (in 1993) was in reality, a secret, agreed upon suspension between Jordan and the NBA as punishment for Jordan's gambling.

The evidence:

1. According to Dave Anderson, Jordan was seen gambling in Atlantic City on Monday, May 24th (despite the fact that he was to play in a playoff game against the New York Knicks the very next night).

2. Jordan has since admitted to compulsive gambling habits due to his competitive nature. (CBS News).

3. Jordan retired from the NBA on October 6, 1993, despite being at the top of his game having just won his third straight NBA championship (and his third straight NBA finals MVP). (Ira Berkow).

4. Richard Esquinas, a San Diego Businessman penned a 1993 book *Michael and Me*, in which Esquinas alleged that he had won close to a million dollars from Jordan from betting on golf.

5. During his retirement from the NBA, Jordan pursued a career in professional baseball. However, he quit after the Arizona Fall League in 1994, despite having a legitimate shot at making it all the way to the Major Leagues with the Chicago White Sox during the 1995 season. (*Sports Illustrated*)

Application of the Bayes Theorem:

h = Michael Jordan's 1993 retirement was in reality a secretly administered suspension from the NBA for gambling.

e = propositions 1–5 listed above.

Notice the way that h seems to make sense of, and entirely explain, all of the seemingly unexpected pieces of evidence. Why would Jordan be in an Atlantic City the night before a finals game? Why would such a competitive person retire at the peak of their career? Why would someone claim Jordan lost over a million dollars betting on golf? All of these pieces of evidence are perfectly explained by the hypothesis in question!

As is the problem with all conspiracy theories, the explanatory power of the hypothesis is high. And yet, the prior probability of the hypothesis, that the NBA functions as a secret society suspending players in secret, is ridiculously low. And, as a result, the ridiculously low prior probability swamps out intuitive appeal that results from the explanatory power of the hypothesis.

John F. Kennedy's Assassination

The theory: Lee Harvey Oswald did not act alone in killing President John F. Kennedy. He was aided by one or more of groups such as the CIA, the Russians, or the US mafia.

The evidence:

1. Jack Ruby, a Dallas nightclub owner with ties to the underworld shot Oswald while he was in custody on November 24th 1963, just two days after Oswald allegedly killed Kennedy. (William Scott Malone).

2. Ruby died just over three years later in prison. He expressed the desire to speak with the Warren Commission but only after being moved to a more secure location as he feared for his own safety in prison in Dallas. (Warren Commission Hearings).

3. The JFK assassination files have still not yet been fully released. The current schedule for release is 2021 (Ian Shapira).

4. Oswald had temporarily defected to the USSR.

Application of the Bayes Theorem:

h = Lee Harvey Oswald's assassination of President John F. Kennedy was in reality a conspiracy that included as co-conspirators some other prominent group (such as the CIA, the Russians, or the US mafia).

e = propositions 1–4 listed above.

Notice the way that h seems to make sense of, and entirely explain, all of the seemingly unexpected pieces of evidence. Why would someone with ties to the underworld murder Lee Harvey Oswald? Why did Oswald's murderer conveniently die in prison after expressing a desire to speak to authorities? If there is no cover-up, why have the details of the investigation not been released? Why was it that someone with Russian sympathies was the one who assassinated a US president? All of these pieces of evidence are perfectly explained by the hypothesis in question!

As with all conspiracy theories, the explanatory power of the hypothesis is high. And yet, the prior probability of the hypothesis, that Lee Harvey Oswald was acting on behalf of a criminal conspiracy, is ridiculously low. And, as a result, the ridiculously low prior probability swamps out intuitive appeal that results from the explanatory power of the hypothesis.

Kurt Cobain's Murder (?)

The theory: Kurt Cobain didn't commit suicide. He was murdered by someone hired by his soon-to-be ex-wife, Courtney Love.

The evidence:

1. Kurt Cobain's alleged suicide letter never in fact claimed he was committing suicide. While the final three lines are fairly cryptic, they appear as if they could be in a different handwriting than the rest of the note. (*Who Killed Kurt Cobain?*)

2. Cobain had an extraordinarily high level of heroin in his system at his time of death. This level seems like it could have been high enough to make him unable to commit the act himself. ("Kurt Cobain's Downward Spiral")

3. The shotgun used to kill Cobain was not checked for prints until over a month after his death. While four latent prints were found. They were deemed unusable. (*Who Killed Kurt Cobain?*)

4. Cobain's attorney, Rosemary Carroll, claimed that Cobain was not suicidal and that he had asked her to draft a new will excluding Courtney Love because he was planning to file for a divorce.

Application of the Bayes Theorem:

h = Kurt Cobain was murdered at the behest of Courtney Love.

e = propositions 1–4 listed above.

Notice the way that h seems to make sense of, and entirely explain, all of the seemingly unexpected pieces of evidence.

Why would Cobain write a suicide letter that is so convoluted? How could Cobain physically accomplish his suicide given the level of heroin in his system? Why were there so many different prints on the gun used to kill Cobain? Why would someone who was not suicidal commit suicide? Isn't it convenient (for Courtney Love) that Cobain died right before excluding Courtney Love from his will?

All of these pieces of evidence are perfectly explained by the hypothesis in question! As with all conspiracy theories, the explanatory power of the hypothesis is high. And yet, the prior probability of the hypothesis, that Courtney Love conspired to murder Kurt Cobain, is ridiculously low. And, as a result, the ridiculously low prior probability swamps out intuitive appeal that results from the explanatory power of the hypothesis.

Adolf Hitler's Ultimate Fate

The theory: As the Russians took Berlin in the Spring of 1945, Adolf Hitler did not commit suicide. Rather, he escaped Europe on a U-boat, sailed to Argentina, and lived out his days in seclusion.

The evidence:

1. Declassified FBI documents contain a number of purported sightings of Hitler after his alleged April 30th death.
   <https://vault.fbi.gov/adolf-hitler/adolf-hitler-part-01-of-04/view

2. After his alleged suicide, Hitler's and Eva Braun's bodies were burned before the advancing Soviet Army could reach them. Skull fragments purported to be those of Hitler (which included a bullet hole) were kept by the USSR (and later, the Russian Federation) for decades. However, a DNA test in 2009 revealed to the fragments to be from a woman under the age of forty. (Andrew Osborn)

3. It has been well documented that a number of high-ranking Nazi officials successfully fled to South America after the war. (Guy Walter)

4. Juan Peron, President of Argentina for nearly a decade following the war was sympathetic to and welcomed Nazis into the country. Eva Peron, Juan's wife, had a bodyguard who was a former Nazi commando. (Peter Crutchley)

Application of the Bayes Theorem:

h = Hitler faked his suicide and fled to Argentina.

e = propositions 1–4 listed above.

Notice the way that h seems to make sense of, and entirely explain, all of the seemingly unexpected pieces of evidence.

Why would so many individuals report sighting of Hitler if he were dead? Why would Hitler's body be desecrated by his own loyal soldiers? Isn't it convenient that so many Nazis successfully fled to South America? Isn't it convenient that Argentina enjoyed such close relationships with so many known ex-Nazis? All of these pieces of evidence are perfectly explained by the hypothesis in question!

As is the problem with all conspiracy theories, the explanatory power of the hypothesis is high. And yet, the prior probability of the hypothesis, that Hitler escaped to Argentina, is ridiculously low. And, as a result, the ridiculously low prior probability swamps out intuitive appeal that results from the explanatory power of the hypothesis.

How to Build a Conspiracy Theory at Home

By now, our recipe for conspiracy should already be apparent. In each of the previous cases, it's alleged that a person or group has operated covertly to bring about some personally advantageous state of affairs. However, what is more important for our purposes is that there is a probablistic structure which underlies conspiracy theories. It is this structure which is at the heart of the recipe. Given the structural similarities that exist in the examples just considered, we can exploit an understanding of Bayes's Theorem and an intentional application of the base rate fallacy to

Step 1 = find some seemingly unrelated bits of information (i.e. some evidence that, by itself, seems entirely random or unconnected).

Note: if done correctly you will have some $P(e)$ that is low.

Step 2 = propose a story that does a perfect job of connecting all of those seemingly unrelated events (i.e., some hypothesis that very nearly entails all the evidence).

Note: if done correctly you will have some $P(e|h)$ that is really high.

Step 3 = emphasize the explanatory power of your story and ignore or obfuscate the reasonableness of the story. The likelihood ratio will be very top heavy and the prior probability of the hypothesis will be ignored.

Note: if done correctly you will draw attention to the $P(e|h)$ and attention away from the $P(h)$.

By drawing attention to the explanatory power of our story and failing to consider the importance of the prior probability of the story, we are ignoring relevant information; we are committing the base rate fallacy. However, the goal in building a conspiracy theory is not to encourage rational thought.

What We've Done

We can understand the intuitive (and appropriate) appeal of conspiracy theories. They offer very good explanations of seemingly surprising evidence. However, we can also correctly identify the epistemological problem with conspiracy theories. The prior plausibility of the story is so low that its explanatory power is relatively insignificant.

We have also tried to illustrate this analysis by appealing to some well-known examples of conspiracy theories. Our analysis, if correct (and let's be honest, it is), should apply to many other situations as well. And hopefully we have learned an important lesson: just because someone is telling a story that sounds plausible given all the dots it's able to connect, we should withhold judgment until we have some sense of the initial plausibility of that story. After all, explanatory power is just one piece of the puzzle!

Finally, we have tried to simplify our analysis into a convenient recipe for building conspiracy theories. As a result, while you should be able to avoid the epistemic blunder of being taken in by a conspiracy theory, you should also be able to exploit the uninformed—hopefully for better, and not worse.

As history has shown, even the most radical of conspiracy theories has some chance of turning out to be true. So, while it would be a mistake to over-estimate the *plausibility* of a conspiracy theory; it would also be a mistake to ignore the *possibility* of a conspiracy theory turning out to be true.

5
How Fallacies Fuel Conspiracies

DAVID KYLE JOHNSON

There are a number of cognitive biases and logical fallacies that help generate belief in conspiracy theories.

Take the Dunning-Kruger effect, for example, people's tendency to (roughly put) believe they are smarter and more capable than they actually are. Ignorance can blind a person to their own ignorance. Since conspiracy theorists consistently think they know better than experts—that they are "in the know" and all others are just "sheeple"—the Dunning-Kruger effect undoubtably plays a role in people starting to believe conspiracy theories.

Or take the representative heuristic which, among other things, causes people to believe that effects should resemble their causes. "Like causes like." In one study, psychologist Patrick Leman showed that people were very unlikely to invoke conspiracies to explain a news story about a gunman who shot and wounded a world leader, but were very likely to invoke a conspiracy to explain the same story if the leader was killed. In reality, of course, the only difference would be where the shooter was pointing his gun. But since people think big events must have big causes, invoking something like the Illuminati seemed necessary to explain why the gun was pointed two inches to the right.

Then there is HADD (our "hyperactive agency detection device") which causes us to conclude that "agents"—persons with intentions and goals—are behind natural events, and apophenia which causes us to see patterns that don't really exist. In concert, these lead to belief in conspiracy theories by causing people to see grandiose patterns in world events that don't exist and then to attribute them to nefarious agents working behind the scenes.

But once a conspiracy theory is established in a person's mind, motivated reasoning is what keeps it going—what fuels it, if you will. And in my experience, there are a few specific cognitive biases and logical fallacies that are especially noteworthy in this regard: confirmation bias, evidence denial, availability error, suppressed evidence, subjective validation, and what I call "the factoid fallacy," "the countless counterfeits fallacy," and "the mystery therefore magic fallacy." It stands to reason, therefore, that if you wish to avoid being duped by conspiracy theories, you need to learn how to recognize these fallacies and how to avoid them. We'll now take a look at some of these fallacious ways of thinking. (For more on many of the fallacies and biases talked about in this chapter, see Steven Novella's *Skeptic's Guide to the Universe*.)

Confirmation Bias

Confirmation bias is the human tendency to only look for evidence that confirms what you want to believe or already think is true. The prime example can be found in many people's cable news viewing habits; they only (and obsessively) watch the channel that tells them what they want to hear. And you can see how prevalent this tendency is in people by looking up how high the ratings are for the news channel that does this best. But confirmation bias fuels conspiracy theories too. Once the conspiracy theorist believes, they will seek out only the evidence that confirms the conspiracy is true; they will never intentionally seek out evidence that it is false. Indeed, this is why some people elect to watch only certain news channels: they feed conspiracy theories.

The tendency to engage in confirmation bias is powerful, and often demonstrated with a game:

> I'm thinking of a number pattern. The sequence 2, 4, 6 follows that pattern. You are supposed to discover the pattern I'm thinking of by giving me more strings of numbers and asking if they fit the pattern. What do you say?

If you are like most, you think the pattern is "ascending even numbers" and try something like "8, 10, 12." That sequence does match, but that's not the pattern I have in mind. Your mistake? You tried to confirm your theory with a sequence that matched it. What you should have done is tried to prove your theory wrong by giving me a sequence that doesn't match it, like 1, 3, 5. When I tell you that does match the pattern, you'll have proved yourself wrong but also be one step closer to the truth. You'll now think the pattern is numbers ascending by two; but don't test that theory with a pattern that fits it, like 7, 9, 11. Try something like 10, 20, 30. When I say that also fits, you are again wrong but closer to the truth. Do you now think it's just any three ascending numbers? Well don't guess "1, 2, 3." Guess "3, 2, 1." Turns out, that *doesn't* fit the pattern—and your theory was right: any three ascending numbers.

This little game also demonstrates how engaging in confirmation bias is one of the most powerful ways to delude yourself. You could have confirmed your "ascending even numbers" hypothesis all day, but never discovered the real pattern. In fact, if you look hard enough, you can find some evidence for anything; but if you don't also seek out contrary evidence, you can fool yourself into thinking that thing true when it's not. Want evidence that Santa is real? Millions of people believe that he is real, have professed to see him personally and say that he delivered their presents on Christmas. There are even dozens of songs about him. If you only sought out those facts, and not any evidence against Santa's existence, you could fool yourself into believing that he exists when he doesn't.

And this is something that fuels conspiracy theories. In their famous study, "Conspiracy Theories: Causes and Cures," Sunstein and Vermeule demonstrated that one way conspiracy theorists engage in confirmation bias online is by combining into online communities, now known as echo chambers, where they are only exposed to information and arguments that confirm their already established conspiratorial views. Of course, there are online communities for sharing information about everything, even spinning tops. But another study (by Del Vicario et. al.) showed that information spreads and is reinforced in conspiracy theory echo chambers differently—and more dangerously. Put simply, unlike factual scientific information which is quickly spread and absorbed by interested parties, conspiracy theories linger and fester, long after they have been debunked.

Evidence Denial

Another upshot of Del Vicario's study was that, in conspiracy theory echo chambers, confirmation bias goes hand in hand with evidence denial. Instead of just *seeking out* only confirming evidence, the conspiracy theorists will also willfully ignore or deny disconfirming evidence when it's presented or happened across. The human propensity to do this, especially with issues that one is passionate about (like conspiracy theories), is well documented. In a study by Brendan Nyhan and Jason Reifler, when conservatives who supported the gulf war were presented with direct evidence that Saddam did not have WMDs, those conservatives more adamantly believed that he did.

Generally, to deny the evidence, people will either explain the evidence away, ignore it entirely, or even try to turn it into evidence for their theory. Explaining the evidence away occurs when "end of the world" predictions fail. Take for example, Harold Camping, who predicted the end of the world in the 1980s and the 1990s. Did the clear evidence of the world not ending convince him that he had no basis for thinking the world was going to end? No! He just explained it away by claiming he had made calculation errors and made another prediction: the rapture would happen on May 21st 2011. And when no one flew up into the air on that day, he again explained the evidence away by saying there was actually a "spiritual" rapture that happened up in heaven that we couldn't see. This might seem funny—and it is—but some of his followers sold all their positions and donated their life savings in anticipation of this event. Later Camping just ignored disconfirming evidence. He had said, after the rapture on May 21st, that Jesus would return on October 21st. When that didn't happen, his radio station just played reruns and he went into hiding for a while. He did eventually admit to having made mistakes, and then subsequently died in 2013.

Some of the best examples of just flat-out ignoring contrary evidence comes in beliefs about government conspiracies. In 2008, the boyfriend of a student (who was obviously distraught by all the critical thinking skills my class had given his girlfriend) came to me arguing that there was going to be a government mandated "bank holiday" in September, to keep everyone from getting their money out of the bank after the government instigated martial law. His claim was based on a rumor going around at the time, as reported by David Mikkelson. I told him I would bet him any amount of money that wouldn't happen. He didn't take the bet. But when it didn't happen, he didn't admit that he was wrong. He just ignored his error, and moved on to the next government conspiracy, and then the next. In 2015, he was trying to convince people that the Jade Helm military exercise was really a front for Obama declaring martial law. That didn't happen either. And multiple failed martial-law conspiracy theories are direct evidence that such theories are bunk; but that fact was simply ignored and he's still endorsing them. (My student eventually dumped him.)

Turning evidence against your theory into evidence for it is very prevalent in conspiratorial thinking because conspiracies essentially have that kind of excuse baked right in. If there's evidence that proves a conspiracy isn't real, the conspiracy theorist can simply insist that evidence was planted by the conspirators to cover their trail—to throw us off the track. "Yeah, that's what *they* want you to think." If, for example, I pointed to the episode of "Adam Ruins Everything" which demonstrates why the parallel shadows in videos of the moon landing were impossible to fake with cinematography in 1969, the "moon landing hoax" conspiracy theorist will simply insist "that's what they want you to think" and conclude the show is in on the hoax too.

To demonstrate the problem with this line of reasoning, consider what I call the ultimate conspiracy theory:

> There are nefarious actors, I know not who, with nefarious intentions, I know not what. All I do know is that all the other conspiracy theories out there were planted by these actors to throw you off the track. So you think the moon landing was a hoax, the Earth is flat, and 9/11 was an inside job? Yeah, that's what they want you to think!

Notice that, by using the logic of conspiratorial thinking itself, the ultimate conspiracy theory proves all conspiracy theories false. So, in a way, conspiratorial thinking is actually self-refuting. But it also demonstrates how irrational denying the evidence is. Nothing could prove this theory false; anything that did would be turned into evidence that it is true. And that's what makes it absurd: it's unfalsifiable.

The Availability Error and Subjective Validation

Something that fuels both confirmation bias and evidence denial is called "the availability error," the tendency to pay attention to or be compelled by the readily available evidence— that which is obvious, noticeable, or what you want to hear— instead of taking into account all the evidence there is. For example, when buying a car, you're more likely to base your decision on the experience of one person you know than on a consumer report that provides statistical evidence (which collates many people's experiences).

The availability error itself is fueled by the fallibility of memory, specifically our tendency to remember things that confirm certain beliefs and forget the things that don't. This is often called "remembering the hits and forgetting the misses" and professional psychics take advantage of it with gusto. They'll make multiple predictions, or say many things that could be true, because probability dictates that at least one of them will be—and then that's the prediction that people will remember. People will forget all the other predictions the psychic made that were wrong which prove that they are not psychic at all.

Another trick psychics use is taking advantage of something known as subjective validation. A claim's truth is validated *objectively* when some kind of objective test can be done, like a measurement. A claim is validated *subjectively*, however, when you decide whether a statement is accurate based on your reaction to it—whether it feels or seems to be true to you. For example, subjective validation is at play when you read the predictions of soothsayer Nostradamus. It may seem, after the fact, that he was writing about World War II, but in reality, his writings are so vague and open to interpretation, they could be about anything (or, in actuality, nothing at all).

A similar tactic is used by astrologers and those who write horoscopes. They will say something vague, and then bet that people will interpret it as applying to themselves specifically. Usually, that bet pays off. In one famous study, Bertram Forer showed that, when presented with the same vague personality profile, around 84 percent of people will think that it describes them. This is now called the Forer Effect.

Availability error and subjective evaluation fuel conspiracy theories in many ways. For example, availability error is at play when conspiracy theorists are apt to notice and remember anything, no matter how small, that confirms their conspiracy of choice, and to not notice or forget anything, no matter how big, that proves it wrong. For example, thanks to Oliver Stone's movie *JFK*, everyone remembers that JFK's head moved "back and to the left," supposedly as it would if shot by someone standing in the grassy knoll (which was in front and to the right of Kennedy). But everyone forgets the rather mundane fact that, when shot, objects usually fall toward the point of impact because of the explosion that the bullet causes when exiting on the other side. Bullets slip into their targets with little resistance, but leave with a lot. Kennedy's head moved back and to the left because the bullet that Oswald shot into the back-left of his head made the front-right of Kennedy's head explode.

Subjective evaluation fuels conspiracies because whether an event or piece of information supports a conspiracy theory is often a matter of interpretation. In the evening sky, you will simply see trails of condensation produced by jet plane engines, which are actually always there, but made visible by the changing angle of the sun light. A conspiracy theorist will see evidence of a government conspiracy to control our mind with "chemtrails." On a trip to the Denver Airport, the conspiracy theorist will see "DZIT DIT GAII," "Au Ag" and dots on a time capsule capstone (donated by the New World Airport Commission) and see clues that members of the New World Order, with the help of an alien race, are going to press the buttons on the capstone to release "Australian antigen" into the atmosphere to dwindle the world's population. You will see the Navajo words for the Colorado Rockies, the chemical symbols for silver and gold (which are mined in the Rockies), and the braille translation of the capstone's inscription (respectively) as pointed out in Brian Dunning's book, *Conspiracies Declassified*.

Suppressed Evidence and the Factoid Fallacy

You commit the suppressed evidence fallacy when you present an argument but leave out evidence or information (either willingly or unwillingly) that would show the conclusion of the argument false. It's obviously related to the topics of the last sections—especially confirmation bias, evidence denial, and the availability error—but is different because it is a mistake committed while making an argument, rather than a human psychological tendency. To clarify this distinction, we might say that confirmation bias, evidence denial, and the availability error often compel us to suppress evidence when making an argument.

Examples of this fallacy abound. Advertisers commit it when they argue that you should buy their product but leave out the evidence that you should not. In 2009, in ads for their wrinkle cream, Olay suppressed the evidence that they retouched the images they used of the sixty-two-year-old actress Twiggy. In arguments against homosexuality, religious fundamentalists will quote Leviticus 20:13, "If a man lies with a male as with a woman, both of them have committed an abomination," but leave out the fact that this prohibition lies in the middle of a list of other prohibitions that they willfully ignore: including the prohibition against consuming pork and shellfish (11:4-12), sex during a woman's period (18:19), wearing mixed fabrics (19:19), beard trimming (19:27), tattoos (19:28), and charging loan interest (35:37).

But it fuels conspiracy theories when it takes the form of what I call "the factoid fallacy," where one takes some true fact that seems to support a view, but leaves out information that explains why it doesn't. For example, it is indeed a fact that jet fuel does not burn hot enough to melt steel. (The hottest jet fuel can burn is about 1500°F; steel doesn't melt until 2750°F.) This fact has been used, ad nauseum, by conspiracy theories to fuel the idea that 9/11 was "an inside job." What they leave out is that steel does not have to be melted—turned into liquid—in order for it to be significantly weakened and unable to support weight. In reality, at the temperature at which jet fuel burns, steel is extremely malleable and is not able to hold up any structure. As David Dunbar points out, steel-framed buildings are brought down, for this very reason, by ordinary fires all the time.

Another perfect example is this factoid that is used to fuel the conspiracy theory that climate change is a hoax: ninety-five percent of all the CO_2 that is produced every year is the result of natural forces. That's true! And, by itself, it makes is seem like climate change could not possibly be the result of human activity. But what those who cite this factoid leave out is the additional fact that natural forces take out just as much CO_2 as they put into the atmosphere; only humans put it in without taking it out. So, humans are responsible for one hundred percent of the *increase* in atmospheric CO_2 each year; natural forces are responsible for none (John Rennie, "Seven Answers to Climate Contrarian Nonsense").

This is why non-experts should listen to experts. The experts know the relevant information and thus whether some factoid should be considered convincing or needs to be put in context. And this is why we should definitely defer to the experts when they are in agreement, like they are with climate change. According to John Cook and his colleagues, a full ninety-seven percent of papers published by experts on the topic agree that climate change is real and caused by humans. If you needed to cross a bridge, and ninety-seven engineers said it would collapse when you tried to cross it, and three said it was fine, you wouldn't cross the bridge.

You may have heard (from conspiracy theorists) that Cook's ninety-seven percent statistic is wrong; but those who argue it is wrong are not experts. They are journalists for *Forbes,* or bloggers for Watts Up With That. When other climate scientists reviewed Cook's work by trying to replicate his analysis, they got the same results. Six different studies have showed that Cook was right, and every attempt to show that he was wrong has been debunked. In fact, when climate scientist Rasmus E. Benestad (et. al) examined the dissenting papers in the three percent, errors in them were found that, when corrected, brought them in line with the consensus view. Conspiracy theorists are merely trying to suppress the evidence for that which is true: climate change is happening, and it's caused by us.

The Countless Counterfeits Fallacy

You commit the counterfeits fallacy when you take a large portion of faulty evidence for a conclusion to be good reason to think that conclusion is true. When stated like that, it seems absurd—so absurd that you might assume that no one ever commits this fallacy. But they do. Indeed, even academics and intellectuals commit this fallacy. In an article about Catholic belief in ghosts, Boston College Philosophy Professor Peter Kreeft argued that the large number of fake and false ghost sightings was evidence that ghosts exist. "The existence of [a great deal of] counterfeit money strongly argues for the existence of real money somewhere." There are so many ghost sightings, his argument goes, that they can't all be fake. And if just one such story is true, ghosts exist.

In conspiracy land, this kind of reasoning is often used to fuel the conspiracy that the government is covering up alien visitations. The fallacy usually appears in an argument like this: First, the conspiracy theorist will pull out the big guns: The Roswell incident, the famous Phoenix lights, and the UFO spotted during a 1991 Mexico City solar eclipse. You, then, debunk all these: Roswell was Project Mogul, the Phoenix lights were leftover flairs from a military exercise, and the "eclipse UFO" was just the planet Venus. But this won't deter the conspiracy theorist.

> Yeah, but there are so many other stories about UFO sightings out there! Hundreds! Thousands! It's unlikely that they are all false, right? And if only one of them is real, aliens exist, so it's likely that aliens are real and the government is covering it up.

Intellectuals commit this version of the fallacy too. I'm almost quoting verbatim a former colleague—who was trained in logic!

It's hard to pinpoint exactly what is wrong with this line of argument. There must obviously be something wrong with it because there's tons of terrible debunked evidence for every conspiracy theory out there, and even for the most devout conspiracy theorist, it would be a stretch to believe them all. But the mistake seems to lie in thinking that whether a piece of evidence is good is a matter of chance so that, the more you find, the more likely it is that one of them is going to be true. While it is true that if you keep throwing dice, you'll eventually get a Yahtzee, that is not how evidence works. Whether a piece of evidence is good is not a matter of chance.

To see why, suppose I lined up a thousand witnesses, that don't even know you, and paid them all to say you murdered someone named Bob to get you convicted of murdering Bob. Now, obviously, this would prove nothing. In each case, the evidence of a paid-off stranger is very weak. But by the line of logic we have been discussing here, I could say, "But there are a thousand of them! What are the chances that they're all wrong? And if only one is right, you're guilty." Obviously, this is not how evidence works.

In fact, the logic behind this line of reasoning is exactly backwards. By simply debunking just a few pieces of bad evidence for something, I actually provide good reason to think that all such evidence is flawed. Consider an example. Suppose a young boy, amazed by his first magic show, sets out to find out whether the performer really has magic powers. In doing so, he figures out how the tricks were done: sleight of hand, misdirection, and illusions. He investigates another magician, and another . . . all turn out to be using simple tricks. After just a few such investigations, isn't the boy justified in believing that all magicians are using tricks—that none of them have supernatural powers? Of course! Even if he came across a new magician, with feats he had never seen, the boy would still be justified in believing that the magician was using simple tricks to fool him. By debunking just a few, he effectively debunks them all.

And so it is with any massive amount of low quality evidence for any conspiracy theory. The fact that there is a lot of terrible evidence for a conspiracy theory is not good reason to think it's true. Indeed, once you have explained away just a few such pieces of evidence, you're justified in just ignoring the rest. In fact, once you have debunked a number of conspiracy theories in this way, you would be justified in doubting any others that come along. (Although, of course, a few mundane conspiracies have actually occurred.)

The Mystery Therefore Magic Fallacy

The magician example from the last section might make you wonder: "What if the boy came across a magician who did things he couldn't explain? Suppose he investigated and came up dry. Wouldn't the boy be justified in believing the magician had magic powers then?" No. And if you thought that, you fell victim to yet another logical fallacy that fuels conspiracy theories: the Mystery Therefore Magic Fallacy.

You commit the Mystery Therefore Magic Fallacy when you take the fact that something can't be explained (that it's a mystery) to be evidence of magic. When committing this fallacy, "magic" need not necessarily be construed as "magical powers" however—although it often is. But it could also be any kind of supernatural, paranormal, or pseudoscientific explanation that the one committing the fallacy wishes to invoke. "I don't know, therefore aliens" . . . or "ghosts," or "Bigfoot," or "miracles." Whatever is a person's favorite explanation will fit. Indeed, that's partly how you know this line of argument is fallacious. The fact that you can't explain a magic trick is just as much evidence that the performer has magic powers, as it is that he has alien technology, or is being helped by a ghost—and those can't all be true together.

Why is this line of reasoning faulty? It's a variety of the appeal to ignorance fallacy, the idea that not being able to prove something false is a reason to think it's true (and vice versa). In this case, not being able to prove that some event isn't supernatural (by finding the natural explanation) is thought to be reason to conclude that the event is indeed supernatural. This is fallacious because you might not be able to prove that it is false for other reasons. Maybe the evidence isn't available. Maybe you are looking in the wrong place. Maybe you just aren't that bright.

Now, technically speaking, appealing to ignorance isn't always fallacious. If you would expect to see evidence of something if it were true, not finding it is evidence that it is false. Not finding any milk in your fridge, for example, is good reason to think that there is no milk in your fridge because you would expect to see it if it were there. And, in general, when it comes to existential matters—questions about whether something exists—the burden of proof is on the believer. A lack of solid evidence for Bigfoot is good reason to think that Bigfoot does not exist. So, we could wonder, might there be an exception for seeming magical events too? If you really look hard for a natural explanation but don't find one, could that be good reason to think there isn't one?

In short, no. Why? Because natural explanations aren't the kind of things that you would always expect to find even if they are there. Consider Penn and Teller, two of the most famous and knowledgeable magicians in the world. They host a show called *Fool Us*, where they invite other magicians to try and do magic tricks that they can't explain. They are fooled at least once an episode, but never once, in being fooled, have they concluded that the participant actually had magical powers. Even though they are likely the best in the world at finding natural explanations for such things, when presented with a performance they can't explain, "We aren't smart enough to figure it out" is still, always, the better explanation. And if that's true for them, the same is true for you. In order for "it's magic" to be the justified conclusion in the case of a mystery, "it's magic" will have to be the best explanation. But in any such case, your own ignorance will always be the better explanation. It's just simpler.

How does this fallacy fuel conspiracy theories among conspiracy theorists? "My favorite conspiracy theory is true" is their magic explanation. Instead of invoking "magical powers" when they come across something they can't explain, they invoke their conspiracy theory of choice. So every mystery becomes another piece of evidence for the conspiracy. "We can't explain the pyramids?" the conspiracy theorist will ask. "Well then it must have been ancient aliens." Of course, we actually have explained the pyramids. We know quite well how they were made, as explained by Benjamin Radford. But even if we didn't, that would not be good reason to invoke conspiracy theories involving ancient aliens. Our own ignorance would be the simpler explanation.

Save Yourself!

Conspiratorial thinking is a rabbit hole; when a person believes one, they often believe many—even if the theories contradict one another. This is likely because the cognitive biases and logical fallacies that fuel conspiracy theories, some of which we have discussed, can lead us astray on any topic. If you don't know how to guard against them, you could end up believing anything. The safety line, to ensure that you don't fall down the conspiracy theory rabbit hole, is to study and know them. Be able to recognize logical fallacies when you see them committed by yourself and others. Be mindful of your cognitive biases and guard against them. Because of things like confirmation bias and evidence denial, pointing out cognitive biases and logical fallacies in the arguments of conspiracy theorists may not always be effective at changing their minds—although, to be honest, you may change the minds of others in the conversation, and recent studies have shown that debunking efforts are not totally ineffective. But at least you can avoid being led astray.

PART II

*"It's who's controlling
the Illuminati that
I'm more concerned
about."*

6
Falling Off the Edge of the Earth

COURTLAND LEWIS AND GREGORY L. BOCK

Flat-Earthers are growing in number and include famous people, such as Kyrie Irving, and enterprising individuals, such as the guy who built and launched his own rocket last year to check out the shape of the Earth for himself (he didn't get high enough).

Flat-Earthers like these believe that there's a conspiracy to hide the true shape of the planet and give various reasons why they think this is true. They receive a lot of ridicule for holding their views, but the one thing that strikes those of us who take the time to listen to their arguments is that they aren't idiots. Most are well-intentioned, good-natured people who are curious about the world. They use evidence and experiments to defend their point of view and have YouTube videos explaining the details. Heck, they even build their own rockets!

But while they may be intelligent, somewhere along the line, their reasoning goes terribly wrong. In this chapter, we explore Flat Earth theory and discuss its flawed epistemology, specifically how it relies too heavily on verification and too little on falsification and how it fails to properly consider information from reliable sources.

In the history of human civilization many people have thought that the earth is flat. Perhaps they even verified this for themselves, but simple verification doesn't equal truth. Just because we can construct a coherent explanation of an observed phenomenon, it doesn't mean it's correct. For a conspiracy theory to be plausible, it would need to take counterevidence seriously, and this is exactly what Flat Earth theories fail to do.

Conspiracy theories are fun because they put all the evidence together in a neat, often fanciful package. They make life

entertaining and meaningful. In a world where most explanations of reality require advanced degrees in mathematics,
flat-earth theories provide answers that all common and
thoughtful people can understand. What's more, and possibly
most importantly, conspiracy theories make us feel significant because they create a community of hope, connectedness,
and intimate relationships based on the belief that "we" have
discovered something nefariously hidden from the rest of
humanity. But is the Earth really flat? Does Flat Earth theory withstand philosophical criticism? Let's see!

Flat Earth Theory

Flat-Earthers believe there's a conspiracy to hide the true
shape of the Earth. They believe that its flatness is known by
the major powers, some shadowy organizations and, of course,
by Flat-Earthers themselves. Beyond that, Flat-Earthers agree
with one another on very little. For the sake of simplicity, we
will take the views of Mark Sargent to be representative of the
lot. Sargent is articulate and intelligent and is currently their
most prominent spokesperson because of his well-known *Flat
Earth Clues* video series on the Internet. The videos sparked a
renaissance in Flat Earth theory, and the movement has since
grown to thousands, maybe hundreds of thousands of followers.

What follows is our own "reader's digest version" of
Sargent's Flat Earth theory. He believes that the world powers
discovered the edge of the Earth in Antarctica around 1956 and
a dome ceiling in the sky around 1958. Upon the Antarctica
discovery, the US government signed the Antarctic Treaty with
several other nations in order to hide the fact, he claims, that
Antarctica is actually an extremely long ice wall. The treaty
makes the continent off-limits until 2041 to all commercial or
civilian interests. After discovering the dome ceiling, both the
US and USSR started shooting high-yield nuclear warheads
into the sky trying to break through. They called these
launches "atmospheric testing" to conceal their real purpose.

To hide the truth from the public, the US government faked
the moon landings and pieced together photos from aircraft to
create photos of the so-called "globe." Astronauts and space
agencies are "in on it" and are sworn to secrecy. The educational system brainwashes all of us from a young age with
models of the globe in every classroom. Hollywood reinforces
the solar system model in its science fiction films. And the FAA
doesn't track flights over the southern oceans because this
would reveal flight paths that make no sense on a globe model.

Why hide the true shape of the Earth? In the fifth installment of the series of clues, Sargent imagines what would happen if the government disclosed the truth. He says it would
bring excitement to the world but would lead to the ending of all
space programs, cause widespread unemployment for astronauts, observatories, astrophysicists, etc. and upend current
power structures in a dramatic fashion. There would also be a
revival of religion, which could be good or bad because religion
could either join with science to work together in understanding
this new world or might fill the vacuum left by the collapse of
the status quo and become a totalitarian power itself. In the
tenth installment of the series, Sargent suggests that science is
trying to hide evidence for God, for if we live under a dome, then
presumably there is a dome-maker. And this revelation would
threaten the secular and atheistic scientific establishment.

What evidence does Sargent provide to support his claim
that the Earth is flat? There's only one line of positive evidence,
and that's personal experience—it appears flat to any casual
observer. The rest of the evidence he presents is negative in
nature in that it focuses on the "flaws," gaps, and anomalies of
the received view. As he says in Part 3 of *Flat Earth Clues*,
"This is what I like to focus on—the gaps, the holes in the plot,
the unanswered questions."

Examples of the reasons Sargent uses to support his claims
include the following. First, there are no 180 degree (or 360
degree!) videos taken in space, which can only mean—Sargent
infers—that NASA is trying to hide the edges of the set.
Second, there are no non-composite photos of the Earth, which
tells us, he claims, that we can't get high enough (above the
dome) to take a photo of the whole thing. Third, southern hemi-
sphere flights don't take the shortest paths to their destina-
tions, if we assume the Earth is a globe. They only do if the
Earth is flat. Fourth, there has never been a movie made about
the moon landing because this would reveal how easy it is to
fake. These are the main reasons provided, but there are more.

Verifying the Truth

A common misconception is that prior to the Enlightenment of
the eighteenth century, people thought the Earth was flat.
Some people in Europe during the Middle Ages thought it was
flat, but ancient Greeks such as Aristotle (384–322 B.C.E.)
believed otherwise. Eclipses of the moon show that the Earth
casts a curved shadow, and stars on the horizon change slightly
depending on the observer's position. Since both of these obser-

vations imply a spherical object, Aristotle concluded the Earth must be spherical. He even calculated its circumference to be 46,000 miles (Thomson and Missner, *On Aristotle*).

With such insights, it might be strange to think that Aristotle has something in common with Flat Earthers. What they share is their method for *verifying* truth. If you're familiar with science and academics, they love to use words in different ways, because language is complex. We use "to verify" in the same sense as "to confirm" or "to corroborate," just as Sargent and other Flat Earthers confirm/corroborate/verify the Earth is flat through their experience.

Often considered the first "scientist," Aristotle combined careful empirical observation with impressive logical reasoning, to categorize and derive several theories and truths about the nature of the world. His teachings were not only influential during his lifetime, but would influence Muslim and Christian thinking, culminating in Scholasticism, the dominate approach to education throughout Europe until the Enlightenment.

Scholasticism combined the theology of Christianity with the science of Aristotle to create a system for drawing conclusions and understanding the nature of the world. Scholasticism dominated Europe until further observations raised doubts about some of its fundamental conclusions, such as the Earth being the center of the universe—more on this later. Even with the collapse of Scholasticism, Aristotle's use of verification to draw conclusions about the world would remain the standard approach to natural philosophy—what would later be called science—until Karl Popper (1902–1994).

To understand how verification works, its strengths, and weaknesses, let's do a thought experiment. Look around at your surroundings and observe what's happening—it might help to go outside or look out a window. Based on your observations, would you say the Earth is at the center of the universe or a planet orbiting the sun at 67,000 miles per hour? You might be tempted to say the latter because you've always been taught that the Earth orbits the sun, but observation suggests otherwise. If you were traveling at 67,000 mph, there would be no humans, trees, or buildings standing. Wind gusts of 60 miles per hour cause enough trouble. Multiply that by a few thousand, and there would be little left on the Earth to be blown away—imagine standing on top of an airplane in mid-flight. What's more, if we were spinning around the sun at such a speed, the centrifugal force would cause us to fly off the planet or be squashed to its surface. Earth would resemble the ball of a roulette wheel. Neither of these events are observed, so empirical evidence suggests we're not orbiting the sun at 67,000 miles per hour. Flat Earthers sometimes use the same thought experiment to support their claims.

Aristotle made the same observation over two thousand years ago and used further empirical evidence to conclude that the Earth must be at the center of the universe. Can you think of an observation that explains why Aristotle arrived at such a conclusion? Think of the roulette wheel. Is there a way to prevent the ball from flying to the side? Maybe it would help to think of your childhood playground. Playgrounds sometimes feature a roundabout (or merry-go-round) that spins in a circle, where kids desperately hang on for their lives while centrifugal forces sling them off. What's the one safe place to be on the roundabout? The center! If you're in the center, centrifugal forces won't affect you. This is an easily verifiable truth, which implies: if we live in a universe with circular orbits, and we're not moving, then we must be in the center of the universe.

Aristotle's understanding of the natural world was the result of what would later become known as verificationism, and verification would remain the standard of empirical knowledge until the mid-1900s. As seen previously, and in the next section, Flat Earthers (and most all conspiracy theories) rely on verification to support their conclusion that the Earth is flat, but there are some serious problems with verification, the main one being that almost any truth-claim can be verified. Think how simple it was for Aristotle to show that the Earth is the center of the universe. To verify something you merely need to state the conclusion you want, then look for evidence that verifies the conclusion. In some instances, verification is sufficient for providing truth. For instance, if you think you left your keys on the kitchen table, you need only examine the kitchen table to verify that's where you left your keys. With more complex truth-claims, such an approach doesn't take into account the complexities involved in understanding the phenomena in question.

Another feature that makes verification problematic is confirmation bias. Simply stated, most humans aren't good at considering and accepting opposing beliefs. Instead, we only look for facts that support our beliefs. If I'm conservative and think liberals are out to take my guns, then anytime I hear someone talk of gun control, I'll think, "There are those liberals again, trying to take my guns." If I'm liberal and want to show that supporters of the president are hate-filled racists, then any person who acts or speaks out of hate or racism will be labeled a president supporter; and if there's no evidence they are a supporter, then it'll be used to show that the president has created an atmosphere where such people are emboldened. If you don't like those examples, imagine you think the Earth is flat. You will look for evidence that verifies your beliefs, and when someone offers you counterevidence, you'll use such evidence as further confirmation for why your belief is true—"Of course, you'd say that, because you're a sheep that has swallowed the government's conspiracy of lies." When such reasoning is used, you've created a nice set of circular beliefs that only allow for other beliefs that are consistent with the rest of your beliefs. As a result, your beliefs fail to count as warranted, for warrant requires not only consistent beliefs, but they must seriously consider counterevidence and match up to the facts of the external world.

Truth and Falsifiability

To fully understand the conceptual issues with only verifying hypotheses, we must understand the nature of science. As mentioned previously, verification was once thought of as the hallmark of science, but is now considered pseudoscience. Instead of verifying our theories, science requires we be able to falsify them—hypothesizing a genuinely risky prediction designed to disprove the hypothesis. In other words, instead of trying to verify, you attempt to falsify your claim by predicting what should or shouldn't happen if your hypothesis were false.

There are several great examples throughout history, but one fun example comes from Edmund Halley. After conducting extensive research on comets, Halley predicted in 1705 that a comet—what we now call Halley's comet—would appear in 1758 in a particular area of the sky, proving that some comets orbit the sun. The truth of his hypothesis rested on a prediction fifty-three years in the future, and at the time seemed highly unlikely. Yet in 1758, sadly after Halley's death, the comet appeared and provided evidence for Halley's hypothesis. If the comet had not appeared, or appeared at a different time and place, then the hypothesis would've been falsified.

Karl Popper was the first to offer a fully-developed conception of falsifiability as the basis of true science—as opposed to pseudoscience. According to Popper, all observations are selective and value-laden, what we discussed earlier as containing confirmation biases. What's more, it's impossible to test a universal proposition. The universal claim "all ravens are black" would require that I verify every raven—past, present, and future—is black, which is impossible. However, due to the nature of universal claims, if I find one counterexample, then it implies the falsity of my claim: if one raven is non-black, then it must be false that all ravens are black.

In terms of a scientific theory, however, one counterexample doesn't make it false. Scientists make predictions and develop tests that attempt to falsify the hypotheses of a theory, and when such tests fail, we can conclude that a theory is strong, or has a high measure of reliability. Finding contradictory truth claims raises doubt about a theory, and suggests either a refinement of the theory or the consideration of a better theory, but doesn't discount the theory as a whole. Just like with ravens, instead of throwing out our understanding of the color of ravens, when we find a white one, we create a sub-category—albino ravens—to allow for the existence of non-black ravens; but if we started finding multi-colored ravens, we would need to redefine what it means to be a raven. Scientific theories are constantly refined as a result of falsification, and in cases where there's a theory that better explains the phenomena of the world, theories are replaced.

The argument that the Earth is the center of the universe was eventually replaced with the heliocentric model that claimed the sun to be the center, because it better explained—it was simpler and less convoluted—the nature of the Earth's rotation in relation to other celestial bodies and several other phenomena observed in our solar system. Other explanations, such as gravity and atmosphere also contributed to the acceptability of the heliocentric model.

Like all scientific inquiry, at the heart of Flat Earth theory is the attempt to solve a problem. To evaluate their arguments, we can't simply examine and accept data and theories that verify the hypothesis that the Earth is flat. We must also look at motives, for they will tell us why individuals accept certain facts, while rejecting similar contradictory facts. All science is value-laden, but real science makes clear and attempts to limit any value assumptions that might taint experiments and conclusions.

We must also determine whether Flat Earthers are simply verifying their hypotheses, or whether they're making risky predictions that would raise doubts about the explanatory force and truth of their position. Let us perform such a task by looking at two examples from the documentary *Behind the Curve*, which illustrate a true scientific attempt to prove the soundness of Flat Earth theory: 1. the ring laser gyroscope (RLG) test, and 2. the laser experiment.

According to the RLG experiment, if the Earth is rotating as the spherical model suggests, then a gyroscope should tilt fifteen degrees after an hour. To obtain the most accurate results possible, two Flat Earth researchers used a RLG and predicted that no tilt would be recorded after an hour. On the video, the RLG registers a fifteen-degree tilt, which bothers the researchers; but like good scientists, they refine their experiment to limit any factors that might result in incorrect data. No matter the refinement, the RLG consistently registered a fifteen-degree tilt. Such results create objectively independent data that contradicts Flat Earth theory. Such results happen often in scientific experimentation, and when they do, scientists must appropriately adjust their theory to be consistent with the external facts of the world—this is the scientific process of falsifiability. To reject external facts of the world, whether you're a traditional scientist or a Flat Earther, is to reject the scientific process in favor of some other subjective desire to retain your belief regardless of the evidence, which is not science and doesn't promote truth.

The second experiment proposed using a laser to test the curvature of the Earth. According to the experimenters, if the Earth is flat, we should be able to place two people at an appropriate distance—one observer and the other pointing a laser towards the observer at what should be the same height—with two screens with holes placed in the middle for the laser to shine through, then when the laser is turned on, the observer should see it. If the Earth is curved, then the observer won't be able to see the laser because the laser-pointer will be below the observer due to the curvature of the Earth.

In *Behind the Curve*, the researchers conduct the experiment, and to their surprise it only works if the laser-pointer is raised high enough to account for curvature, thereby, contradicting the prediction that the Earth is flat. This is a perfect example of the process of science—gathering data and offering falsifiable predictions. Scientists deal with negative results every day, but the process of science calls for researchers to ignore their motives and let scientific reasoning rule. To hold on to a theory in the light of unbiased, objective data is to be dogmatic, not scientific. Again, one example doesn't prove a theory false, but scientific data must be repeatable, and to ignore strong evidence that contradicts your theory, for whatever reason, undermines the truth of your theory.

Think for Yourself

Even if Flat Earthers begin to show more concern for falsification when they test their hypotheses, there's still something epistemically distressing in the way they deal with counterevidence. For example, if you suggest to Flat Earthers like Sargent that we now have many non-composite photos of Earth from deep space, such as from the Deep Space Climate Observatory, they reply that this information comes from NASA, which is a government entity and can't be trusted. If asked why NASA can't be trusted, they argue that the moon landing was faked. And when you press them on this claim, they produce many well-worn claims based on anomalies in the photographs, such as the waving flag on the moon, no blast crater under the lunar lander, and the letter "C" on a moon rock.

The best explanation for these oddities, they argue, is that we didn't go to the moon at all, but rather the US faked the moon landing in a movie studio with the help of someone like Stanley Kubrick in order to win the space race—never mind the number of actors who would have to be "in on it" and would have to be trusted to take the secret to their graves, even though blowing the whistle would likely net them instant fame and fortune. This doesn't even address the fact that the USSR, who had the most to lose if NASA "cheated," didn't contest the fact that the landing happened. Anyway, with regard to the anomalies in the moon-landing photographs, there are plenty of plausible alternative explanations that can be Googled, which don't appeal to conspiracy theories.

The skepticism of Flat Earthers isn't limited to space programs. Dig a little deeper and you discover an infectious distrust of any institution—public or private, government or scientific. There may be many psychological reasons for this, but our purpose here isn't to psychoanalyze our Flat Earth friends but only to argue why such distrust is unjustified. And the best way to do this is to simply examine the structure of these institutions. Take scientific institutions like universities, for example. Scientists gain employment at universities by earning advanced degrees in their field and producing quality work that is published in peer-reviewed journals. Failure to publish results in failure to gain employment or loss of a job (for those on tenure track). The peer-review process is difficult and competitive and isn't a matter of just "toeing the party line" or regurgitating "dogmas." It's about creativity and critical thinking. It's about proposing hypotheses and testing them rigorously. Peer reviewers aren't members of a secret society in a secret room somewhere; they're other professionals in the field, thousands of them across the world who scrutinize submitted work for originality and methodological rigor.

The Flat Earth mantra is "Think for yourself," and while this sounds excellent at first, it's disguising something pernicious. Thinking for oneself, what's known as the virtue of intellectual autonomy, is indeed, an epistemic virtue, but like all virtues, it's a mean between two extremes. On the one side, there's the vice of gullibility, which means blindly believing whatever the authorities tell you. On the other side, there's the vice of epistemic isolation, which means cutting yourself off from the world or having your head in the sand. Isolation is blind ignorance, and while blindness in both senses ought to be avoided, the Flat Earth mantra leans more toward blind ignorance.

The virtue of intellectual autonomy isn't opposed to trusting the testimony of others and experts. If it were, we could hardly know anything at all and our situation would be dangerously close to solipsism or Descartes' *Cogito*—trapped in our own limited experiences of the world. Part of acquiring knowledge is learning what sources of information to trust. For example, I've never been to New Zealand. Can I trust that there is such a place? I've seen maps, pictures, and even *Lord of the Rings*, but unless I can trust the mapmakers, photographers, and film directors, I'm not justified in believing there is such a place.

But that's absurd! There's no reason to disbelieve in the existence of New Zealand, and maps, photographs, and movies can count as evidence in favor of its existence. In regard to the conclusions of scientific experts, unless we're in possession of good reasons to doubt them, then we're justified in believing what they say about their areas of study, even about cosmology and the shape of the Earth. This doesn't constitute blind trust because we can always withdraw our assent and double-check claims that cause us concern. Being autonomous and engaging in knowledge acquisition involves trust, and this is natural and good.

Things Left Unsaid

Understanding the world in which we live is difficult. Understanding science and how it describes the world is even more difficult. We've tried to make sense of some of the Flat-Earth arguments, how they arrive at their conclusions and how they ultimately get the scientific method wrong, but there are several arguments we haven't addressed. For example, there are many Flat-Earthers who are motivated by their interpretations of religious texts. For them, truth is, at least partly, a matter of faith in scripture.

As Mark Sargent notes in some of his videos, there's an interesting correlation between some world religions and Flat

Earth theory. But for these religiously-informed claims about the natural world to be taken seriously, their scriptural interpretations should be open to counterevidence and modification, especially if they accept the fruits of the scientific method in other areas, such as contemporary medicine and engineering. To accept science's view of the natural world in one area and ignore it in another seems to be inconsistent and more about protecting one's cherished beliefs than about discovering truth.

Though well-intentioned and good-natured, Flat Earthers approach evidence and science incorrectly. Instead of supporting their conclusions through careful scientific analysis, they undercut their position by relying on pseudoscience. Without true scientific experimentation, there are no good reasons to accept their conclusions, except that they are more fun or personally meaningful than the scientific explanations of reality.

If you're a Flat Earther, you'll think we're simply delusional or operatives of the government—we wish! We're good critical thinkers who want to know the truth but aren't willing to abandon the means for obtaining truth in order to feel better about our place in the world, whether that's spiraling through the universe at 67,000 mph or sitting comfortably on a disk at its center.

7
That's . . . One Giant Joke on Mankind?

Jeff Cervantez

"The Eagle has landed," Neil Armstrong said. His voice was heard crackling across the airwaves in NASA's mission control center in Houston. The date was July 20th 1969 and it was reported across the globe that Armstrong had successfully landed the Apollo 11 lunar module, dubbed the "Eagle," on the moon.

Shortly thereafter, Armstrong was seen emerging from the lunar module. At the same time nearly a billion people were watching and listening to these events. In bewilderment, they witnessed Armstrong step onto the moon's surface and deliver his now famous words, "That's one small step for a man, one giant leap for mankind."

In the coming years, several more Apollo astronauts would return to the moon. The last moon-landing was in December 1972. Between July 1969 and December 1972 a total of twelve astronauts would step onto the moon, the final being Gene Cernan during the Apollo 17 mission.

Most of the world celebrated these historic events, but some came to question whether they ever even happened. From the time of the Apollo missions to the present day, people have doubted whether the moon-landings ever took place. They insist that the moon-landing missions were only a clever hoax. For most of human history the possibility of such an undertaking was fantasy, completely impossible. And so, it's just hard for some people to believe that human beings were capable of this accomplishment, especially in the year 1969. This seems to be the impetus for what I call the moon-landing hoax.

The moon-landing hoax is the claim that the six Apollo moon-landing missions between 1969 and 1972 were all faked. For one reason or another, the deception was orchestrated by

NASA, with the help of the US government to intentionally mislead people into thinking humans had reached the moon. Somewhat surprisingly, this belief is not relegated to dark pockets of the Internet. In 2018, the two-time NBA Most Valuable Player Stephen Curry made headlines when he questioned the moon-landings, thus joining the ranks of moon-landing skeptics. Curry is by no means alone in his doubts. Some recent opinion polls in both Britain and Russia have revealed that a sizable percentage of the public is skeptical about the American moon landings. Opinion polls in the US have also found that many Americans express doubts about the veracity of the moon-landings. There seems to be a critical mass of people who persist in believing that the moon-landings were a fraud.

One of the most notorious moon-landing skeptics is the self-proclaimed conspiracy theorist, Bart Sibrel. Sibrel is known for publishing books and producing movies investigating the Apollo missions. He too claims the moon-landings were a hoax. Sibrel has gone so far as to publicly challenge Buzz Aldrin on the subject. Aldrin was Armstrong's fellow astronaut on Apollo 11. In an interesting exchange that you can see on YouTube, Sibrel is seen approaching Aldrin saying, "You're the one who said you walked on the moon when you didn't." Aldrin is heard saying, "Will you get him away from me?" Sibrel proceeds to call Aldrin "a coward, and a liar, and a thief." Aldrin can't take it anymore. The seventy-two-year-old unloads with a solid right-hand punch to Sibrel's face, nearly taking him down.

So, you might wonder, when it comes to the theories about the moon-landings, is exchanging insults and blows the best we can hope for? Fortunately, we can do better. A little philosophy can help us settle this dispute without the need for anymore name calling or right hooks.

At issue are two alternative theories about the moon-landings. The first is not so much a theory, but rather the claim that the moon-landings are a historical fact. It's the belief that the moon-landings actually happened. Let's call this theory the standard view of the moon-landing, or simply the standard view.

The alternative to the standard view is that the moon-landings were a hoax. This view is a direct challenge to the standard view. It asserts that the historical facts about the moon-landing are wrong. This alternative account claims that the evidence against the standard view is enough to undermine its veracity. More than that, this alternative theory states that the standard view was intentionally fabricated so as to deceive the public.

Evaluating Competing Theories

In evaluating these respective theories, I'm going to follow a four-step process.

1. **Clearly state the theory in question.**

2. **Examine the evidence for the theory.**

3. **Consider a plausible alternative theory.**

4. **Determine which theory is the simplest, yet most reasonable, explanation of events.**

I've adapted these four steps from *How to Think about Weird Things* by Theodore Schick, Jr., and Lewis Vaughn. As the authors point out, this is an exercise in applied epistemology. Epistemology is the philosopher's big ten-dollar word for the study of knowledge. A question in epistemology might be: How can know that a particular factual claim is true? Or, is that theory justified?

In step one of our process, we need to clearly state the idea of the moon-landing hoax theory. If you've been following along so far, then you have a fairly good idea of what the hoax theory is claiming. It contends that the standard view of the moon-landing is a fraud. The rationale usually given for this shocking amount of duplicity is that the US government was motivated to win the cold war. The objective was to trick the Russians into spending vast sums of money on their lunar program thereby destabilizing the Soviet economy and triggering the downfall of their government. That's the claim. Now what's the evidence for the theory?

There is no *direct* evidence for the hoax theory. It is a conjecture based on what we know about the political climate at the time. To be sure, the US and SovietRussia were locked in a cold war during the space race. So, a deception on the scale of a moon-landing hoax is logically possible. The question is, is the hoax theory likely?

To this point, no US documents or American leaders or NASA employees have emerged to corroborate this idea. Consequently, instead of providing direct evidence to support their theory, proponents of the moon-landing hoax usually point to a laundry list of circumstantial evidence. The evidence is a list of claimed discrepancies between the available facts and the standard view. These discrepancies are supposed to show that the moon-landing hoax is the best explanation of the available evidence. Is the hoax theory successful in accomplishing this objective?

In exploring this question, we must look at the list of so-called inconsistencies with the standard view. This list varies but usually includes some common lines of reasoning. After putting these points on the table, we are then in a position to juxtapose the moon-landing hoax and the standard view.

So, what's the evidence for the moon-landing hoax theory? Here's some of the strongest evidence given in defense of this theory.

- The American flag that was placed on the moon by Apollo 11 appears to be blowing in the wind, but there is no wind on the moon.

- In the photographs taken on the lunar surface there is an absence of stars in the dark horizon.

- Some of the images from the lunar surface contain curious shadows. The shadows seem inconsistent with the lunar surface but appear more consistent to what you would find in a staged studio with artificial lighting.

- The high temperatures and radiation found on the lunar surface would have destroyed the film used by Apollo astronauts.

- No one can pass safely through the Van Allen belts that surround the Earth. These lethal belts of radiation preclude safe travel.

These points are curious and fascinating. But, do they support the rather stupendous claim that they are supposed to support? Should these lines of reasoning lead to the conclusion that the moon-landing was a hoax? Do they undermine the credibility of the standard view to the point that we need a radically different theory of events?

I don't think so. Astronauts and experts in the scientific community have provided what look like reasonable explanations for these anomalies.

- The flag appears to be blowing in the wind because pieces of wire were weaved into the top and bottom of the flag to help it keep its shape when planted on the moon. NASA wanted the flag to appear extended, as if blowing. This would allow the flag better visibility in photographs.

- The absence of stars in the photographs is due to how NASA set the cameras. In order to capture images in a bright lunar environment, the settings are such that dim sources of light are drowned out on the film.

- The curious shadows are solely due to the viewer's perspective when looking at an object. This has been demonstrated on Earth and the same principles apply on the moon.

- With respect to the heat and radiation on the moon, the astronauts, and their cameras, were covered with helpful reflective materials that limited the amount of heat and radiation that would be absorbed. And most of the astronaut's activities were completed during lunar dawn, thus they did not experience high temperatures.

- It's true that the radiation of the Van Allen belts is lethal, but only if astronauts were to spend several days in their radioactive vicinity. Astronauts typically pass through these belts within an hour. Consequently, their exposure to radiation is no more dangerous than that from receiving a couple of CT scans or X-rays.

What does this all mean for the moon-landing hoax theory? The hoax theory is on shaky ground. To persist in accepting the hoax theory without reservations at this point would make us guilty of confirmation bias. This is because even absent alternative theories, the moon-landing hoax theory is now highly questionable. Reasonable counter-evidence has been provided to explain the discrepancies which seemed to support the hoax theory. It would take a large logical leap to continue embracing the hoax theory in the face of a reasonable alternative. When we confront powerful contrary evidence to our theory, it is sensible to begin exploring possible alternative theories. Fortunately, we needn't look very far for a plausible alternative.

Contrary to the hoax theory, the standard view claims that the moon-landings are a historical fact. The moon-landings happened more or less according to NASA's historical record, as certified and verified by the larger scientific community.

The evidence for the standard view is wide-ranging. Reviewing all the evidence here would be far too lengthy a process. It is simply not possible to list all the possible corroborations for the standard view within this limited space. Fortunately, Neil Armstrong has provided us with a cogent summary of the clearest and most logical evidence in defense of the standard view.

In the years following Apollo 11, Armstrong was periodically questioned about the veracity of the moon-landings. In the biography of Armstrong titled *First Man* by James R. Hansen, Hansen records Armstrong's response to such inquiries. In response to critics of the standard view Armstrong would often reply that the lunar flights "are undisputed in the scientific and technical worlds. All of the reputable scientific societies affirm the flights and their results." Furthermore, "the crews were observed to enter their space craft in Florida and observed to be recovered in the Pacific Ocean. The flights were tracked by radars in a number of countries throughout their flight to the moon and back. The crew sent television pictures of the voyage including flying over the lunar landscape and on the surface—pictures of lunar scenes previously unknown and now confirmed. The crew retuned samples from the lunar surface including some minerals never found on Earth."

Ockham's Razor

At this point in the evaluation process we need to determine which theory is the simplest, yet most reasonable, explanation of events. Does the moon-landing hoax provide the clearest and most straight forward account? Or, does the standard view provide the better understanding?

The medieval philosopher William of Ockham (1285–1349) has left us a valuable tool for addressing these questions. Ockham said, "Never increase, beyond what is necessary, the number of entities required to explain anything." In other words, keep your explanations simple by cutting away all unnecessary assumptions. Because of the emphasis on cutting away speculations and assumptions from theories, this logical principle has become known as Ockham's Razor.

Quite simply, Ockham's Razor is a philosophical criterion for deciding between competing theories. All things being equal, we should prefer the simpler theory—the one that makes the smallest number of assumptions. Ockham's Razor is also known as the principle of simplicity or the principle of parsimony.

Ockham's Razor raises a revealing question for our discussion. Which theory makes the minimum number of assumptions that are required for explaining the presented phenomena? According to Ockham's principle, we should always prefer the simpler explanation. When deciding between completing theories we should favor the one that requires the fewest assumptions while still accounting for all the known phenomena.

Applying Ockham's Razor to the competing accounts of the moon-landing we can ask, Which theory is simplest? By "simplest" I mean which theory proposes the least number of logical steps in their account of the facts? It should be clear why the standard view of the moon-landing is preferable at this point. One of the biggest challenges for the hoax theory is that it makes a lot of assumptions. It presents an elaborate, and sometimes complicated, picture of the facts.

For example, briefly consider the number of people who would need to be complicit in a moon-landing hoax. If the hoax is going to work, the thousands of people working at NASA and many others working for the US government at the time of the Apollo missions would have to be involved in a deep cover-up. These same people would have to continue to deceive other people for years, without ever telling another person of their deception. If the moon-landings were in fact a hoax, it's difficult to explain why not one person with knowledge of the fraud has yet to come forward with the truth.

As the authors of *Weird Things* point out, in order to accept the moon-landing hoax theory we must accept a very complicated and improbable story. Rather than accepting the strong evidence in favor of the standard view, we must insist that "that such a massive and complex conspiracy could be orchestrated without being exposed by whistle-blowers, disgruntled employees, or conspirators; that the thousands of contractors who worked for NASA would never come forward to cry foul; that after more than thirty-five years, no definitive evidence—no genuine documents, files, recordings, or anything else—would surface to reveal the truth; that of all the scientists and engineers from all over the world who have had access to staggering amounts of Apollo data, none would suspect fraud and tell all." It is far simpler, and reasonable, to accept the strong evidence in favor of the standard view.

The standard view could be wrong. It's logically possible that the moon-landings were a hoax. The hoax theory, however, is not the best available interpretation of the facts. As we have seen, the many unnecessary and dubious assumptions associated with the hoax theory, make it a highly undesirable and improbable explanation.

Perhaps this is why Armstrong was reported to have said, "The only thing more difficult to achieve than the lunar flights would be to successfully fake them."

8
Witch Hunts

ALEXANDER E. HOOKE

> A secret and a conspiracy can be effective weapons in the hands of someone who doesn't believe them.
>
> —UMBERTO ECO

Conspiracy theories invite entertainment and curiosity.

The three wise men visiting the baby Jesus were actually aliens from Mars. They were guided not by a brilliant star but by the fires of a mushroom cloud from the world's first atom bomb, as evidenced by an unnaturally humungous crater near Jerusalem. 9/11 was part of US government plot—did you see the photo of a stoic George Bush that morning? He did not look surprised at all. Those murders of Marilyn Monroe (no suicide or overdose), JFK, and John Lennon were orchestrated by government agencies. Lee Harvey Oswald was a CIA stooge and the Chapman kid was under an FBI spell when he shot the ex-Beatle.

One of my favorites from Nick Redfern's *Secret History*: Today's billionaires have hired a secret medical research team so they can live forever while the rest of us die off in misery. Such reports help us laugh off those occasional but all-too-human moments of cosmic paranoia.

One of the most infamous cases of conspiracy theory, though, was actually quite intimidating with the potential to ruin people's lives. These were the witch conspiracies erupting in Europe and New England from the fourteenth to the early eighteenth centuries. The result was the persecution of thousands (some estimates claim more than a half-million) of individuals accused of being corrupted by the Devil. The punishments meted out to these alleged satanic miscreants were quite real—imprisonment, torture, exile, heavy fines and, more often than we care to admit, execution by drowning, hanging or being burned alive in a public setting for all the townspeople to witness.

It is tempting to dismiss the witch hunt phenomenon as a brief setback in the progress of human justice. If we consider some of its central features—accusation based on rumors, condemnation instead of a fair hearing, guilty verdicts regardless of contradictory testimonies and flimsy evidence such as inborn scarification marks inscribed by the Devil upon the accused's body—then we might consider how the attitudes and practices of witch hunt conspiracies still thrive today.

From college students and faculty vilifying guest speakers as racists when they have not even read the individuals' writings, to political tyrants who decry every opposition as a secret cabal that conjures fake news, all the way to rigged events and intellectual hoaxes to undermine the tyrant's devotion to his people, the temptation to carry out witch hunts lingers on the horizons of the human landscape.

Umberto Eco's sardonic epigraph above provides a litmus test for studying several philosophical aspects of the witch hunt as conspiracy. First is the issue of veracity: evidence, legal status, witnesses, motive, contrary testimonies, sources of a confession (with or without torture), a suspect's past and possible relation to the accuser, the accused and accuser's reputation in the community.

A second aspect involves the actual experiences comprising the accusation. Looking at the accused face-to-face in a village courtroom, displaying fits and anguish every time glances are exchanged, insisting that the woman (usually) we have all known for so many years is the culprit for a herd of cows catching a fatal disease, husbands wandering about her house during the night, children suddenly suffering from a strange illness, all stirred existential moments among the denizens. In addition, as folklore often shows, the accused have been seen flying on broomsticks, mixing toxic potions, while deploying black cats and live toads for their mischief and cavorting with Satan's helpers in the dark forests.

In a sense there are two sorts of conspiracies going on with witch hunts. There is the charge of a conspiracy against strange human beings. There is also the use of conspiracy as a weapon by the accuser whose essential task is to persuade others to believe what he or she might not believe, but for the accuser's own benefit.

To study these two aspects of conspiracy, I will rely on the ideas of French philosopher Michel Foucault. Foucault was quite adept in scrutinizing historical contexts to identify how forms of knowledge intertwined with exercises of power. His genealogical accounts of emerging ideals, momentary transgressions, shifts in how humans determine what is true or good, as well as the historical dynamics of the confession, provide a conceptual lens to study how philosophical issues arise in witch hunts. These issues include who determines the truth of accusations of witchcraft, the manner in which a defendant asserts her or his innocence, whether persecution and elimination of the alleged miscreant promises or undermines a happier community, or the extent that divine insight and human judgment reflect or conflict with one another.

Scholars have anchored their studies of witch hunts from a variety of perspectives. Discussed here include: a history of patriarchal capitalist oppression of women; omnipresence of the Devil; women's knowledge of how to mix herbs and plants for medicinal purposes; the on-going human tendency to find and punish a scapegoat; and, social tactics for advancing one's own benefits at the expense of strangers and oddballs.

There is a rational thread to the witch-hunt conspiracy. Citizens really did believe or were persuaded that other people practiced witchcraft. This rational thread could be driven by popular fear and neighborly grudges while encouraged and justified by intellectual figures such as Cotton Mather. His preaching and writing contributed judicial and religious legitimacy to the persecutions. In sum, there is an enduring core to conspiracy theories—they are fueled by excessive passion while guided by an explanatory justification. That can be a formidable combination to prevail over the protestations of innocence made by innumerable young women.

Tests and Trials

It seems that only humans subject each other to tests and trials. That is, only humans present obstacles and challenges to one another that have nothing directly to do with procreation, survival, food or enjoying life. Our responses to a test are couched in an ostensibly rational activity—assessment, ranking and evaluation. We have spectacular weekends for athletic tests (such as the Super Bowl), decades of schoolroom tests in the name of pedagogy, public and televised trials to show how well we obey the law or remain loyal to the cause, as well as inquisitions that attempt to prove who we really are—or are not.

The tests and trials involving witch hunts embody this human legacy remarkably well. The conditions under which accused witches passed or failed provide us with considerable resources for understanding the rationality and absurdity of many conspiracy theories.

Risks of a Witches' Brew

In a sense, there have always been witches, according to Penny Le Courteu and Jay Burreson in their book, *Napoleon's Buttons*. Witchcraft was for centuries considered a variation of sorcery, which meant little more than learning to use or control nature for human well-being. Witches and sorcerers tried to learn and apply the secrets of how plants such as mandrake, belladonna, coca leaves, or meadowsweet tree could improve health or relieve pain. Today they would be called amateur herbalists, as they lacked the scientific approach to these mixtures while always tinkering with new varieties and blends.

Their curiosity was consistent with the original traditions of medicine that stressed the principle of relying on the "botanical basis" of pharmaceutical treatments. Eventually some combinations stirred by witches created new combinations of molecules called alkaloids. One version was ergot. It could be fatal to livestock, and if swallowed by humans could generate "convulsions, seizures, diarrhea, manic behavior . . . vomiting, twitching, a crawling sensation on the skin . . ." Other witches' combinations were likened to miracle cures, for they would produce physical recovery, emotional tranquility, or even spiritual ecstasy.

If doctors could not help them, many patients suffering from an illness or lingering pain sought the counsel of witches. Here is where the molecules come in. While the witches were not specialists in a scientific laboratory, Le Courteu and Burreson describe how inventive these women could be when mixing herbs, roots, leaves, flowers, and seeds. Often their potions really alleviated another person's ailments, loss of appetite or inability to sleep. Still, the authors point out, there were unexpected downsides. With lack of rigid controls, these "witches' brews" could produce some startling surprises. Most prominently, some of the mixtures sparked weird illnesses as well as hallucinogenic experiences that led to visions of deities or devils, not to mention seeing witches flying on their brooms amid the shadows of a full moon.

Around 1500 A.D., the Church's tolerance of benign witchcraft quickly reversed. The Inquisition led to an expansion of the category of heretics, with witches now included under the

Church's ever-expanding scope for God's secret enemies. Soon wild rumors about witches' deeds included orgies, infant cannibalism, sex with demons, deeds "beyond rationality but were still fervently believed." According to Le Courteu and Burreson, an odd twist ensues: "In medieval Europe the very same women who were persecuted kept alive the important knowledge of medicinal plants, as did native people in other parts of the world" (pp. 225, 245).

A Woman's Test

Silvia Federici, in *Witches, Witch-hunting, and Women* contends that the history of women being targeted as witches is central to the tradition of patriarchal capitalist societies. This on-going hunt often served to legitimize a claim to the property belonging to a suspicious woman, lead to public shaming of her character, or threatened her with exile. This witch-hunt tradition, according to Federici, begins with Plato and the early Christian Fathers and extends to the current violence thrust upon women in all corners of the planet.

Central to this demonization of women was the accusation of malicious gossip. Rather than appreciating the communicative lubricant force of gossip as Sissela Bok does in her work *Secrets*, witch hunters often saw women's gossip as purveying poisonous and contagious aspersions on respectable citizens. Ultimately, though, the belief about women's sexuality underscored the most fearful suspicions of witchcraft. Federici notes, "The fear of women's uncontrolled sexuality explains the popularity in the demonologies of the myth of Circe, the legendary enchantress who by her magical arts transformed the men lusting after her into animals."

While historians generally agree that witch hunts in Europe and New England ceased in the early eighteenth century, Federici cites numerous reports of witch hunts going on more recently throughout the world. In Africa, for example, women in parts of Ghana and Kenya are still subject to banishment, enslavement, torture, or execution. She detects a dark "logic of witch-hunting" that underscores this incessant and harsh treatment of women—it rewards the mercenaries hired to capture them, it benefits those who can confiscate the accused witch's property, and it satisfies grudges and vengeances among adversarial tribes.

Curiously, assessing blame by Federici seems to envision still another conspiracy. She requests that a trial be established for all those agencies who have established the conditions for these persecutions while pretending they do not exist. For her, these include "African governments, . . . the World Bank, The International Monetary Fund, and their international supporters—the U.S, Canada, the European Union" (p. 62).

Tests, Trials, and Tortures

Nearly every student taking an introductory philosophy course learns about the trial of Socrates. Most teachers (myself included) tend to adopt Plato's rendition that shows Socrates to be a victim of an injustice in his noble pursuit of truth. The case can and has been made, however, that Socrates invited his arrest with his refusal to compromise his mission as presented by the Delphic Oracle: Socrates must abide by the axioms "know thyself" and "the unexamined life is not worth living" through his encounters and dialogues with friends and fellow Athenians.

The dispute over Socrates's guilt or innocence, according to James Franklin in his *The Science of Conjecture,* was resolved by appealing to the more persuasive argument. It was an adversarial struggle, according to Platonists, between the clever Sophists (alleged masters of rhetorical sleight-of-hand and linguistic tricks) and the devotees of truth (friends and advocates of the Socratic mission). However scholars and students consider the merits or flaws of the case against Socrates, his trial did not involve a test.

Trials for witchcraft relied heavily on the test. Franklin examines what counted as reliable testimony and the probability of accurate evidence for juridical deliberations. While there have always been debates about the trustworthiness of eyewitnesses, material circumstances, and previous histories of the accused in deciding guilt or innocence, a peculiar legal component was introduced during the 1500s. To be accused of a crime would now be considered a partial proof of guilt. This meant partial punishment could be deployed to ascertain the degree of wrong-doing by the accused. In other words, torture became central to eliciting a possible confession from the accused.

The confession was more than admitting to breaking a law. According to Foucault in his lectures (IV and V) on the relation of torture to confession, the trial was a public drama to determine if the accused not only admits to the deed but also confesses to who she really is, that she has made a compact with the devil, that her demonic self seeks the ruination of her fellow humans. All those attending this torture as public spectacle had to witness and judge—with the joy or horror that accompany such spectators' experiences—the sincerity and authenticity of the confession. So, the trial of torture revolved around the accused's ability to withstand the unbearable suffering or relent and confess: Yes, I really am a witch, with the devil in my soul.

If you or I were subject to pins piercing through our throats or lips, rocks piled upon our stripped bodies, burning rods thrust into our loins, most of us would confess to any crime simply to stop the pain. Federici's historical account overlooks this phenomenon—the idea that accusation of a crime implied partial guilt was part of juridical proceedings for all suspects. Next to heretics and blasphemers, witches happened to be easy targets. Franklin recounts one episode where the accused witch was so courageous to withstand the pains and so persistent in her claim to innocence, that the court found her not guilty. It did request that she leave the country. This did not satisfy the mob, who stoned her to death as she tried to depart.

There were considerable objections to the witch trials. Montaigne ridiculed them as flim-flams, but so clever as to dare to "cause a man to be roasted alive." A German Jesuit, Friedrich von Spee, published a mockery of how "evidence" was contrived in witch trials. If the accused can withstand the pain of torture, then she is strong from Satan's support. If she confesses, then it shows that Satan has betrayed her. Circular reasoning was widespread. Pascal once noted that humans are so mad that not to be mad was another form of madness. Such a paradox applied to those accused of witchcraft—to prove their innocence was another satanic trick that revealed their guilt.

Seeing What No One Else Sees

The idle amusement sparked by conspiracy theorists lies in confronting remote facts and unfamiliar sources that catch us off-guard, as if someone else sees what we're oblivious to. This moment tantalizes us with a thought experiment. Suppose the facts are right and there really is a secret cabal, Illuminati legion, or an inner circle of billionaires who control most worldly events. Such facts can only underscore the credibility of the charges.

One essential component of conspiracy theories comes from the intellectual legitimacy accorded to those facts by a leading thinker or writer. In seventeenth-century New England that figure was Cotton Mather. He was a widely respected minister, speaker, and self-proclaimed expert in detecting the Lord's enemies that dared to ruin human life. In his treatise *The Wonders of the Invisible World,* Mather describes the wonders of devils, demons, evil angels, malefactors and witchcrafts that elude us more ordinary individuals. When Mather sees or hears reports about neighbors having seizures, friends bewitched by a strange specter, acquaintances succumbing to a heretofore unknown sickness, he detects not misfortune or bad luck but the secretive workings of the devil. The rest of us are unfamiliar of these workings. This is a consistent attitude of most conspiracy theorists: They see what you and I do not, and cannot without their guidance.

Mather goes one step further. In part II of his treatise, discussing the trial of accused witch Bridget Bishop, he condones treating the accused with various tactics to urge a confession, including choking, pinching, drowning, biting, and related tortures. Regarding the beheadings and hangings of so many women, Mather observed, ". . . if the Storm of Justice do now fall only on the Heads of those Guilty Witches and Wretches which have defiled our Land, How Happy!" (p. 248).

Historians such as Emerson Baker and Stacy Schiff see the conspiracy of witches more as a conspiracy of the accusers, led by mob mentality and legitimized by such credible and Biblically literate individuals as Cotton Mather. Like any other society, early New England had its fill of hopes and disappointments, moments of genuine community and violent forays of conflict and mutual suspicion. Many infants died before childhood; bad weather caused all sorts of havoc with food and livestock; there was on-going tension with American Indians and incoming migrants. The villagers searched for those who might have brought about their misfortunes. In a word, they looked for scapegoats to account for the accuser's or the town's misfortunes. (For those sympathetic to Plato, Socrates is one of the original scapegoats.)

In his *The Devil of the Great Island,* Baker recounts the phenomenon of lithobolia. It is the sporadic exercise of throwing stones at other peoples residences, work places or common meeting areas. Obviously, no one wants stones being thrown at them in the middle of the night. Baker's research of court documents and memoirs shows that much of this lithobolia was a manifestation of local disputes, property disagreements, old scores being settled.

With the support of learned scholars like Cotton Mather and mobs seeking to find the culprit, it was decided that the stone-throwers were encouraged by the devil with the assistance of New England's own furtive witches. Villages and towns in early New England anticipated on-going conflicts with Native

Americans, other colonialists, godless immigrants, . . . not much different from Hobbes's account of the state of nature. When communities were unable to resolve threats or disputes, they assumed a much more powerful force was at work—the work of witchcraft. Its practitioners were outsiders, oddballs, suspicious neighbors, apparent malcontents who seemed to cast malevolent deeds upon others. For the fearful residents of New England, writes Baker, "These outsiders focused and made tangible fears. They also provided scapegoats."

Schiff depicts Puritan culture in the late seventeenth century as a veritable panoptic society. Parents, neighbors, and churchgoers were on the constant look-out for the lustful yearnings of young boys and girls, the unexpected moments of hysteria by young women and impulsive distractions of young men when reaching puberty, or threats to family order as could be engineered by the master deluder Satan. Encouraged by the sermons and writings of Cotton Mather, the Puritan panoptic focus was on detecting the tricks and subtle seductions of Satan.

Here the odd or eccentric woman became a familiar culprit. Second to idolatry, witchcraft was the most important capital crime. "If any man or woman be a witch, that is, hath or consulteth with a familiar spirit, they shall be put to death." Schiff includes original drawings and sketches that portray the alleged witch flying on a broomstick, mixing brew with toads, or lustfully cavorting with Satan's assistants. Amid this furor, Schiff illuminates how fearful citizens turned their anger and suspicions onto the accused in such a coercive way that it forced the accused to ask themselves: "Could I be a witch and not know it?"

Echoing Foucault's theme, this paradox of self-knowledge is fundamental to conspiracies, whether fanciful or quite real. The accused were often placed under such duress—from hysterical witnesses, legal religious authorities, to bodily torture—that it was difficult to avoid providing a false confession, particularly if the accused witch took seriously the threat that the accusers would first punish her children as potential witches. After the weight of the testimony, many would confess to witchcraft. Before the night of their executions, most of the condemned silently reconsidered the events and became convinced again that they were not witches. Thus the next day, approaching the scene of their demise, the condemned witches continually shouted out their innocence to all the spectators eager to watch the hanging or the burning at the stakes.

The Real Conspirers

Nancy Mairs, poet and essayist, once wrote a piece "On Being a Cripple." She had long struggled with a multiple sclerosis, but soon tired of hearing about being "momentarily disabled" or "physically challenged" and related multisyllabic euphemisms. No, she insisted, she is a cripple because that is what the disease does—it cripples you.

Her verb-noun distinction applies to conspiracy theories. Before there is a conspiracy, there must be those who are conspiring. The word derives from "together + breathe." This implies that any alleged conspiracy is first anchored to an encounter where a variety of individuals meet face-to-face to discuss, gossip, speculate over their possible enemies or threats to the common good. Amid closer quarters in a tavern or someone's home, in a metaphorical sense they hear one another's "breathing" that we associate with anger, revenge, misgiving, anticipation of recompense for one's misfortune. Once a target is agreed upon, the attendants then conspire about tactics and strategies, from legal prosecution to public shaming, if not vigilante-like justice of a late night hanging or torture.

Though throughout the history of witchcraft there have been small and reclusive groups who practiced esoteric rituals, eccentric readings and meditations, and secret efforts to use nature's resources to address human suffering and illness, the witch hunts here were primarily focused on isolated individuals who had little recourse to defend themselves. It was an inherently lopsided confrontation. Most importantly, the accused witches were almost never engaged with one another. They did not "breathe together" to cast their spells and Satanic curses upon hapless neighbors and unsuspecting villagers.

In her research, Schiff spotted one or two possible cases of witches conspiring demonic plots. By current juridical procedures, the evidence would be laughed out of the court. At that time prosecutors elicited the wildest testimony of seeing witches meeting in a dark field, exchanging bread and wine as if they were consuming blood. The number of witches seen at this cabal ranged from 24 to 500. Regardless of the inconsistent numbers, jurors would enlist the testimony of a girl who claims she saw the accused to have "suckled birds, hairless kittens, pigs . . ." Many of the accused witches were poor servant girls who were subject to harsh treatment from lascivious swineherds, masters who sexually abused them, and house mistresses who often beat them for disobedience. Should they resist or try to escape, the "witch" charge quickly arose. Elsewhere,

Schiff recounts how prosecutors once relied on the testimony of children who saw a bunch of witches using cats and birds to drink the blood of other children. Later the young witnesses admitted that they had lied.

Umberto Eco points out how conspiracy theories thrive on finding patterns and connections that are coincidental, happenstance, or simply unverifiable. But there is more than this epistemological obstacle. In the case of witch hunts, for example, there was more than just a misuse of logic. There was also a passion fueled by anger or envy in finding a scapegoat. The actual cabal of conspirers were those sundry individuals who collected their resources in secret locations, breathed together their suspicions in order to condemn and eliminate their scapegoat—the witch.

At this point, conspiracy theories shift from informative entertainment to another tragic and human absurdity.

9
Racial Genocide Theories

ROD CARVETH

Conspiracy theories are commonly defined as explanatory beliefs about a group of actors who collude in secret to reach malevolent goals.

One of the most notorious conspiracy theories was pushed by a group of conservative critics known as the "Birthers." Birthers claimed that government officials and journalists lied about US President Barack Obama's birthplace to cover up his supposed ineligibility to be president. This conspiracy theory was advanced by various conservative organs and then by members of the Hillary Clinton campaign against Obama's candidacy, then taken up by the Republican candidate Donald Trump, in addition to radio and television talk-show hosts Rush Limbaugh and Sean Hannity.

Many of the birthers suggested that the time in his youth that Obama lived in Indonesia was proof that he was brought up as a "radical" Muslim. Thus, the explanation for the conspiracy was that Obama was secretly a Muslim, and would seek policies favorable to Muslims while president.

Conspiracy discourse has been a mainstay in US politics—not only among politicians, but among citizens as well—at both ends of the political ideological spectrum. The birther example highlights merely a tiny fraction of the wide range of conspiracy theories that characterize the way many people in the United States think about use or abuse of governing power.

One type of conspiracy theory that has links to the concept of an abusive government is the *race-related conspiracy theory*. In terms of race-related conspiracy theories, there is a difference between a *race-neutral conspiracy theory* (the moon landing was a hoax) and a *race-relevant conspiracy theory* (cocaine was introduced into the black community by the U.S. government).

Thus, race-relevant conspiracy theories involve specific ill intentions toward particular ethnic groups or their leaders.

Perhaps the most powerful and provocative race-relevant conspiracy theories are *white genocide theory* and *black genocide theory*. Both share a common thread that outside forces (the government, the elites, the power structure) are creating conditions that will ultimately lead to the elimination of those of white European heritage or those who are African-American.

Beyond that, the two theories diverge. White genocide theory has been around for about forty years, having gained more prominence in the last decade. Black genocide theory has nearly a century-long history. Research also shows that African Americans in the United States endorse conspiracy theories at greater rates than whites because of a motivation to blame the social system for prejudice and discrimination. The two theories also diverge in that the response to black genocide theory has been mostly political, while, in recent years, the response to white genocide theory has been deadly.

Black Genocide Theory

Candace Owens is a twenty-nine-year-old African American female who is both communications director for the conservative Turning Point USA group, and a founder of the Blexit ("Black Exit") movement. The goal of Blexit is to convert African Americans from being Democrats to being Republicans.

Owens often uses rhetoric such as claiming that African Americans are still "on the plantation" because of Democratic Party policies that leave them dependent on the government. In addition, Owens has charged that the "abortion industry" has targeted the African American community. She claims that despite being less than seven percent of the population, African American women have forty percent of the abortions in the United States. Owens further asserts that nineteen million African-American babies have been aborted since 1972, meaning that the black population would be "double what it is today."

Despite the fact that Owens exaggerates the statistics (African-American abortions are about thirteen million. Add that to the current African-American population of forty-five million and you would get fifty-eight million, not ninety million), the notion that an "abortion industry" is targeting African Americans taps into what is known as "black genocide" theory.

As far back as 1934, Marcus Garvey advocated banning the use of all birth control for African Americans, a position held by many African American organizations, such as the National Urban League (NUL) and the National Association for the Advancement of Colored People (NAACP), well into the 1960s. Even Jesse Jackson, in 1973, claimed that "abortion is genocide."

Anti-abortion activism has attempted to link abortion to racism for decades. The collaboration between conservative black churches, black anti-abortion activists, and some white anti-abortion organizations have promoted the argument that abortion poses a unique threat to black lives, a threat that has seen an increase in attention in recent years. Not only have these activists decried the immorality of abortion, they have promoted the dubious assertion that those who have had an abortion are at five times as much risk to suffer from breast cancer.

Race-relevant conspiracy theories often contain a kernel of truth. There are innumerable historical examples of majority groups conspiring to do harm to minorities. The eugenics movement in the early twentieth century encouraged the use of birth control among poor and ethnic populations to curb reproduction by the "unfit." These policies fueled later rumors that birth control is a form of African American genocide, driving subsequent reductions in the use of birth control in African American communities. Efforts by white reformers only served to reinforce the notion that contraception was a method to assist black genocide.

In 1939, Margaret Sanger and other white birth-control reformers started what was known as The Negro Project. Despite the fact that African-American community leaders were only consulted after the fact, the project did have the backing of such major African-American figures as W.E.B. Dubois and Mary McLeod Bethune. The Negro Project was designed to be an example of a New Deal progressive welfare program. Unfortunately, the program was influenced by the eugenics movement.

It is not clear if what The Negro Project evolved into was what Sanger initially had in mind. She stated that "We believe birth control knowledge brought to this group, is the most direct, constructive aid that can be given them to improve their immediate situation." In other words, Sanger viewed The Negro Project as way to assist African-Americans getting access to safe contraception and maintain birth-control services in their community in the same way she had helped provide birth control access to African Americans in Harlem.

The Negro Project focused its efforts in the rural South, which not only had high birth rates, but was an area suffering from abject poverty. The South was seen as a prime area to demonstrate that contraceptive clinics were essential in impoverished Southern communities. If successful there, the lessons learned could be applied to other areas across the United States. The project focused on African-Americans because they were the group with the most economic and health problems— a group characterized as largely illiterate and that "still breed carelessly and disastrously" according to a report by W.E.B. Dubois himself.

Sanger had proposed that before setting up the clinics, that a period of time be devoted to an educational program to get potential clients to buy in to the program. She envisioned that this educational outreach campaign be led by African American preachers and doctors. In a statement that would color her legacy afterwards, Sanger stated, "We do not want word to go out that we want to exterminate the Negro population and the minister is the man who can straighten out that idea if it ever occurs to any of their more rebellious members." In the end, not only did the project not use Sanger's plan to have African-American leaders to conduct the outreach campaign, but Sanger's words were used to portray her as someone in favor of black genocide through birth control.

It did not help Sanger that another project that took place in the South, the notorious Tuskegee Study, occurred at the same time. In 1932, nearly four hundred African-American sharecroppers who were infected with syphilis were recruited for a "medical treatment" study. In reality, they were not provided with treatment at all, even when penicillin was discovered to be an effective treatment for syphilis. The study continued for forty years, until a news exposé revealed the nature of the project. By that time, about thirty percent of the study participants had died of the disease. More importantly, ten percent of their spouses had contracted syphilis as well as five percent of their children. A similar study occurred in Guatemala beginning in 1946.

These two studies helped drive later suspicions among ethnic minorities that the AIDS virus was created by the government to wipe them out. When AIDS was discovered in the early 1980s, there was a great deal of confusion about the disease. Nobody knew where the epidemic came from. Complicating the matter was the fact that ideas about how AIDS was transmitted, who was at risk, and even what to call it (it had been called Gay-Related Immune Disease) kept changing over the first few years.

By the mid-1980s, medical research was able to determine that HIV was the cause of AIDS. But, the question as to where HIV came from and how it had spread so quickly remained unanswered. There were theories that HIV had natural origins, but there was also speculation that HIV came from contaminated polio vaccines or was accidentally "triggered" by smallpox vaccinations. The most conspiratorial of these theories suggested that HIV was purposely developed to attack groups of people.

As a result, it's not a surprise that the conspiracy theory that HIV was created as a method of genocide in the black community emerged. In fact, by 1991, only four percent of the respondents in a national survey agreed with the statement "the government is using AIDS to kill off minority groups," twenty percent of black respondents agreed with the statement. Unfortunately, subsequent research revealed that those African Americans who were most likely to believe that conspiracy theory were also less likely to engage in behaviors that would make them safer from contracting HIV, such as using condoms. Thus, believing that AIDS was a government plot to put African Americans at risk actually put African Americans at risk.

White Genocide Theory

In April 2018, President Donald Trump tweeted, "There is a Revolution going on in California. Soooo many Sanctuary areas want OUT of this ridiculous, crime infested & breeding concept."

Many critics charge that Trump's tweet was a "dog-whistle" for the conspiracy theory that minorities' higher birthrates are designed to bring about "white genocide" in America. Trump's reference to "breeding" is shared by a number of comments that you will see online about the perils on immigration: "We can't let the Muslims and Mexicans outbreed us"; "Their goal is to come here and breed whites out of existence"; "What they will do is breed like rats and displace white people and white culture"; or "They're outbreeding us and outvoting us."

The white genocide conspiracy theory is a message heavily pushed by white nationalists, who claim that there is a systematic global conspiracy to wipe out the white race through immigration, a concept called "the great replacement." The concept first originated in a 1973 novel entitled *Les Camps des Saints* ("The Camp of the Saints") a futuristic French novel in which France is overrun by hundreds of thousands of migrants from the Indian subcontinent, and, by the next day, France has fallen to "black and brown" invasion. The novel sold poorly on its initial release. In 2011, however, the novel was re-released and found an audience among the right wing in French politics.

The next year, Renaud Camus, a prolific writer and critic, wrote a book entitled *Le Grand Replacement* that argued native "white" Europeans were in danger of being eliminated— in power, if not in terms of absolute numbers—by black and brown immigrants, who were flooding the Continent. As Camus said, "The great replacement is very simple." He has stated, "You have one people, and in the space of a generation you have a different people."

In Europe, the fear is that the growing number of Muslim immigrants will eventually result in white genocide. Even worse, supporters of this theory believe, the policies of the European Union are supporting this move to white genocide. This white genocide theory helped contribute to the vote by Great Britain to leave the European Union in 2016 with its "Brexit" (British exit) vote.

Camus's notion of the Great Replacement has been spread by right-wing and white nationalist figures across the world. In July 2018, Lauren Southern, a Canadian alt-right figure posted, a video titled "The Great Replacement" on YouTube that got over 250,000 views. There is a website called great-replacement.com that declares, "Of all the different races of people on this planet, only the European races are facing the possibility of extinction in a relatively near future." Mike Cernovich, who is part of the white nationalist "alt-right" movement, has claimed on his Twitter account that "diversity is code for white genocide." White nationalist Congressperson Steve King (R-Ia) has echoed his support of "The Great Replacement," and once stated that twenty-five Americans die daily because of undocumented immigrants, and argued that the deaths were "a slow-motion Holocaust." Perhaps the clearest articulation of the great replacement theory is the chant of the "alt-right" marchers in Charlottesville, Virginia, in July 2017 who chanted, "You will not replace us" and "Jews will not replace us."

A further example of white genocide theory occurred in October 2018 when President Trump and the right-wing media virtually obsessed on the notion of an immigrant "caravan" that was making its way up through Mexico as an "invasion." In addition, the caravan was allegedly funded by Jewish philanthropist George Soros.

Soros, a Hungarian-American Holocaust survivor, former hedge fund manager and billionaire Democratic donor, is a long-time target for right-wingers. For example, a tweet that displayed a well-known fake meme that falsely identified a picture of SS member Oskar Groning as a young George Soros was retweeted over six thousand times, yet Twitter took five days after it was reported to remove the post and suspend the poster. The damage had already been done. Soros's critics claim that he not only financially underwrites left-wing protesters, but they he provides the funding to support "invasions" non-white immigrants and refugees. Such views are shared by far-right politicians like Hungarian Prime Minister Viktor Orban and Polish ruling party leader Jarosław Kaczyński.

This "Soros connection" got a great deal of play on the conservative news network, Fox News. For example, on November 13th 2018, Ami Horowitz, a guest on Fox's *Tucker Carlson Tonight*, speculated that "Soros is part" of the caravan, noting that "this whole thing cost millions and millions of dollars." Fox Business host Lou Dobbs had guest Chris Farrell of Judicial Watch on his program who claimed that the refugee march was under the influence of the "Soros-occupied State Department." A "Soros-occupied State Department" conjures up the concept of a "Zionist Occupied Government," or ZOG, a common theme in antisemitic conspiracy theories. *Fox and Friends* broadcast an interview in which Judicial Watch head Tom Fitton falsely charged that Guatemala had become a "way station" for terrorists.

Three days later, NRATV (the online network of the National Rifle Association) correspondent Chuck Holton claimed it was pretty clear "to find a pretty direct link between George Soros money and the people in the caravan getting fed." On the white supremacist message board Stormfront, a user advanced the theory that illegal immigration is a Jewish plot to murder "innocent White Children."

Implications

The problem with most conspiracy theories is that they are often built on a false premise, and, thus, not true. But, when treated as true, and when spread by a number of "trusted" sources, the implications of the conspiracy theories can be deadly.

In mid-October 2018, Cesar Sayoc, a right-wing Trump supporter, sent fourteen packages containing mail bombs to news media organizations such as CNN and Trump critics, including George Soros. On social media, Sayoc posted messages supportive of genocide theory, in terms of Hispanic immigrants as well as Muslim ones. Sayoc accused George Soros of controlling Florida gubernatorial candidate Andrew Gillum, who was potentially Florida's first African-American governor. Sayoc also claimed that Soros was financially supporting David Hogg, a survivor of the February mass school shooting in Parkland, Florida, who both Sayoc and conspiracy theorists like Alex Jones believe to be a "crisis actor." Making the issue worse is that after the bombing attempts, some right-wing pundits like Rush Limbaugh and Ann Coulter jumped straight for more conspiracy, calling the bombs "false flags"—covert operations conducted by governments, corporations, or other organizations, which are designed to deceive the public in such a way that the operations appear as if they are being carried out by other entities.

Days later, Robert Gregory Bowers opened fire on the Squirrel Hill synagogue in the Pittsburgh suburb of Squirrel Hill, killing eleven and injuring seven. Before he left on his deadly mission, Bowers posted on Gab, a social media platform favored by neo-Nazis: "HIAS [Hebrew Immigrant Aid Society] likes to bring invaders in that kill our people." In addition, Bowers made frequent posts that George Soros was funding the "caravan" of Honduran refugees coming to the US seeking asylum. Unline Sayoc, Bowers was not a major Trump supporter, not because he did not sympathize with much of the president's conspiratorial rhetoric, but because he believed Trump's administration was not antisemitic enough.

The same week, a Twitter account called @InvasionPlot appeared posting names and photos and names of Jewish scholars, journalists, student activists, and public officials and listed their pro-immigrant and pro-refugee views. The practice of publishing names on social media is called "doxing"—a practice that has often resulted in people so identified being harassed in real life. The Twitter account garnered thousands of followers before it was suspended.

Thus, though white genocide theory sprung forth from a work of fiction, with no basis in reality, there have been instances of deadly outcomes associated with the theory. Ironically, black genocide theory has at least some connection to actual harms inflicted upon African Americans. Yet, black genocide theory has primarily led its adherents to engage in political behavior, such as switching political parties, or supporting legislation that would make abortions harder to get. Whatever harm came to African Americans from believing black genocide theory was self-inflicted (such as putting themselves more at risk for AIDS).

Both white genocide theory and black genocide theory are conspiracy theories based on largely false assumptions. Yet, the impacts of believing these theories have vastly different outcomes.

10
Do Climate Skeptics Have to Be Conspiracy Theorists?

Don Fallis

There is overwhelming agreement among climate scientists that the planet is getting hotter, that humans are causing it, and that it will have devastating consequences.

The accumulation of carbon dioxide in the atmosphere will lead to more destructive storms, more wildfires, more droughts, and more flooding. And if that's not frightening enough, climate scientists are now projecting that global warming will lead to a serious disruption in the supply of beer!

If the climate scientists are right that all of this is actually happening, we'd better do something about it. But given the degree to which the global economy relies on fossil fuels, significantly reducing carbon emissions is going to be an extremely expensive proposition (if it is possible at all). It's not a step to be taken lightly. So, we have to ask ourselves, should we put our trust in these climate scientists and believe that *anthropogenic climate change* is really happening?

As California burns and Florida sinks into the sea, evidence for anthropogenic climate change that everyone can appreciate is beginning to accumulate. The last five years have been the hottest on record. Scandinavians have started to open wineries. And glaciers clearly haven't been doing so well in Facebook's Ten-Year Challenge.

Even so, it can still get cold enough in the winter for Senator Inhofe to bring a snowball into the United States Capitol in order to show how crazy anthropogenic climate change is. So, how do we know that the evidence of warming is not just due to the natural variability of the weather? And even if the planet is getting hotter, how do we know that humans are causing it, rather than, say, solar activity?

David Hume on Climate Change

In *An Enquiry Concerning Human Understanding*, the renowned Scottish philosopher David Hume (1711–1776) warned us to be skeptical about claims that do not fit with our experience of the world. For instance, he is famously dubious "when anyone tells me, that he saw a dead man restored to life." After all, such a thing "has never been observed in any age or country."

But of course, this doesn't mean that we should never believe something unless we have seen it with our own eyes. It just means that we may need a lot of evidence in some cases. For instance, Hume writes:

> The Indian prince, who refused to believe the first relations concerning the effects of frost, reasoned justly; and it naturally required very strong testimony to engage his assent to facts, that arose from a state of nature, with which he was unacquainted, and which bore so little analogy to those events, of which he had had constant and uniform experience.

In other words, someone who has never seen water freeze is legitimately going to take some convincing that this can happen. Similarly, someone who hasn't experienced the average global temperature increasing by several degrees may also take some convincing that this is going to happen.

Hume did say that there are some things, such as the dead coming back to life, that we shouldn't believe no matter how reliable our informant is. A *miracle* is an event that would require a violation of the laws of nature. In other words, it is something that *conflicts with our best scientific knowledge* about the world. And it certainly wouldn't require a violation of the laws of nature for an informant, even an extremely reliable one, to be mistaken. So, when it comes to the walking dead, just ask yourself, is it "more probable, that this person should either deceive or be deceived, or that the fact, which he relates, should really have happened"?

However, Hume's skepticism about miracles does not support *climate skepticism*. (And it certainly doesn't support climate *deniers* who, unlike skeptics, won't keep an open mind and go where the evidence leads.) Although an inexorably warming planet may not fit with our past experience, it does not conflict with our best scientific knowledge about the world. In fact, as the contemporary philosopher of science Eric Winsberg explains in his *Philosophy and Climate Science*, anthropogenic climate change is part of our best science.

Scientific Consensus

Hume actually provides us with a rationale for believing in anthropogenic climate change. He pointed out that "we entertain a suspicion concerning any matter of fact, when the witnesses contradict each other; when they are but few, or of a doubtful character." But none of these warning signs are present in the case of climate science. A whole lot of climate scientists have reached the conclusion that anthropogenic climate change is true. Indeed, pretty much all climate scientists are in agreement. And like all professional scientists, climate scientists have "such credit and reputation in the eyes of mankind, as to have a great deal to lose in case of their being detected in any falsehood."

Such scientific consensus is compelling evidence that a claim is true. Whenever a large number of people in a position to know something make the very same claim, it is typically very good evidence that the claim is true. It is incredibly unlikely that so many experts would be mistaken in the very same way. But despite the scientific consensus on the issue, many people (including somewhere between a third and a half of the U.S. population) nevertheless deny that anthropogenic climate change is real.

As the American philosopher of science Thomas Kuhn (1922–1996) pointed out in his landmark work *The Structure of Scientific Revolutions*, scientific consensus has been wrong before. In the words of Agent K from *Men In Black*, "Fifteen hundred years ago, everybody knew that the Earth was the center of the universe. Five hundred years ago, everybody knew that the Earth was flat. . . . Imagine what you'll know tomorrow." So, it's not crazy to worry that scientific consensus might be incorrect this time as well. But if anthropogenic climate change *is* false, how do climate skeptics account for the fact that the overwhelming majority of climate scientists think that it's true?

A Vast Global Conspiracy

Much like round earth skeptics (a.k.a. flat-earthers) and other pseudoscientists, most climate skeptics explain away this unwelcome scientific consensus by claiming that there is a *vast global conspiracy*. That is, they contend that "global warming is a myth concocted by scientists," that anthropogenic climate change is "a hoax perpetrated by corrupt scientists who wish to spend more taxpayer money on climate research."

Lack of evidence of a conspiracy is, conveniently enough, evidence for many conspiracy theorists that there *is* a conspiracy. But climate skeptics can actually point to some evidence that suggests that scientists might be working together to fake the data for anthropogenic climate change. The so-called "Climategate" emails, which were stolen and made public in 2009, do seem a little incriminating. For example, Phil Jones, a climate scientist at the University of East Anglia, wrote to his colleagues:

> I've just completed Mike's Nature trick of adding in the real temps to each series for the last 20 years (i.e., from 1981 onwards) and from 1961 for Keith's to hide the decline.

Many climate skeptics maintain that this shows climate scientists colluding in order to conceal the fact that the planet is actually getting colder.

But with only this meager evidence for collusion, maybe we can just reject this sort of conspiratorial thinking. Indeed, the eminent Austrian-British philosopher Sir Karl Popper (1902–1994) famously claimed in *The Open Society and Its Enemies* that it is irrational to believe in conspiracy theories. However, several contemporary philosophers, such as Brian Keeley and David Coady, have recently argued to the contrary that conspiracy theories are not necessarily crazy. Sometimes, they're just where the evidence takes us. Indeed, there have been some actual conspiracies to mislead the public, even in science. For instance, as reported in the bestselling historical work *Merchants of Doubt*, many scientists were paid by the tobacco industry to cast doubt on the scientific consensus that smoking causes cancer.

The English philosopher Thomas Hobbes (1588–1679) even thought that modern science started out as a conspiracy. He accused the *natural philosophers* (a.k.a. scientists) of the Royal Society, including Sir Isaac Newton and Robert Boyle, of being a cabal. For instance, in *Dialogus Physicus de Natura Aeris*, he writes:

> They display new machines, to show their vacuum and trifling wonders, in the way that they behave who deal in exotic animals, which are not to be seen without payment. All of them are my enemies.

Of course, this might just have been sour grapes because they never invited Hobbes to join.

A Vast Global Conspiracy Revisited

But even if conspiracy theories are not always crazy, a conspiracy of the sort that climate skeptics require seems rather implausible. Far too many people would have to be involved. Admittedly, it's not quite as farfetched as the conspiracy that Round Earth skeptics have to believe in. In order to keep it secret that the Earth is really flat, astronauts, airplane pilots, air traffic controllers, cartographers, and many others would all need to be colluding with scientists. In order to keep it secret that the Earth is not getting hotter, just the climate scientists would have to be in on it. But that is still thousands of scientists representing dozens of different scientific disciplines, such as "climatology, meteorology, atmospheric dynamics, atmospheric physics, atmospheric chemistry, solar physics, historical climatology, geophysics, geochemistry, geology, soil science, oceanography, glaciology, paleoclimatology, ecology, biogeography, biochemistry, computer science, mathematical and numerical modeling, statistics, time series analysis, and more."

In addition, the one piece of evidence for a vast global conspiracy of climate scientists (those incriminating emails) has been debunked. As the Harvard philosopher Willard Van Orman Quine (1908–2000) pointed out years ago, our data about the world often have to be corrected based on what we know about the various instruments that were used to collect that data. And this applies even to something as seemingly simple and mundane as measuring temperature. It's not like we've been using the very same, consistently reliable instrument to make temperature measurements for the last millennium, or even the last hundred years.

For instance, in order to estimate average global temperatures before reliable thermometers existed, climate scientists use "climate proxies" like ice cores and tree rings. But since some tree-ring data doesn't coincide with the readings of all of our other instruments (include other proxies) over the last few decades, this data needs to be corrected, which is all that Phil Jones and his colleagues were up to.

But in order to reject climate skepticism, it is not sufficient to undercut their conspiracy theory. A vast global conspiracy is just one possible way to account for why almost all climate scientists agree that humans are causing global warming even if we aren't. In other words, climate skeptics don't have to be conspiracy theorists.

Scientific Consensus without Collusion

Just like anybody else, scientists are subject to the psychological phenomenon of *groupthink*. As Winsberg points out, "to the extent that no climate lab wants to be the oddball on the block, there is significant pressure to tune one's model to the crowd." Indeed, he suggests that this sort of *herd mentality* is nothing new in science. For example, in the early twentieth century, physicists consistently underestimated the speed of light despite using various different techniques to measure it.

It would be extremely unlikely for thousands of scientists to independently reach the wrong conclusion, or for thousands of scientists to be involved in a vast global conspiracy. But it wouldn't be so strange for a handful of scientists to mistakenly conclude that the planet is getting hotter, or for a few eminent scientists to be paid off by Greenpeace, and for everybody else to simply follow along.

The famous "Conformity Experiments" performed by the Polish-American social psychologist Solomon Asch (1907–1996) show that people will even say what they *know* to be false in order not to stand out. Hopefully, scientists are better than that. But groupthink can skew scientific results even if no one is outright lying.

In addition to groupthink, *unconscious bias* is another possible explanation for why almost all climate scientists agree that humans are causing global warming even if we aren't. Indeed, Hume also says that "we entertain a suspicion concerning any matter of fact, when the witnesses . . . have an interest in what they affirm." And this is part of Kuhn's explanation for why scientific consensuses that are incorrect can persist for so long. Even it if is actually false, scientists can usually be productive by continuing to work within a well-established scientific *paradigm*. Thus, they have a motivation to ignore anomalous results, at least until things finally reach a crisis point (and we get a *paradigm shift*).

In his opinion piece "Follow The (Climate Change) Money," Stephen Moore, an economist and one of President Trump's advisors, emphasizes that the lure of money is a possible source of bias in the case of climate science. He writes:

> One reason so many hundreds of scientists are persuaded that the sky is falling is that they are paid handsomely to do so. . . . This tsunami of government money distorts science in hidden ways that even the scientists who are corrupted often don't appreciate.

And, as Winsberg admits, while climate simulations are based on well understood and well confirmed science, they include many parameters that are simply "tuned" to fit the existing data. This provides an opening for the biases of climate scientists to creep in to their results.

Overturning Scientific Consensus

Given that there are other possible explanations for why almost all climate scientists agree that humans are causing global warming even if we aren't, it's not necessary for climate skeptics to appeal to outright collusion. At the end of the day, however, these non-conspiratorial explanations are not that much more plausible than the conspiratorial explanation.

In his opinion piece, Moore goes on to advise young climate scientists that "you're probably not going to do your career any good or get famous by publishing research that the crisis isn't happening." But this doesn't seem right at all. While there are incentives in terms of scientific reputation and financial reward for climate scientists to just go along with the consensus on anthropogenic climate change, there are the same sorts of incentives for climate scientists to try to overturn the consensus. In particular, if you could show that anthropogenic climate change is false, it would make your career. (And if anthropogenic climate change is false, it should certainly be possible to do this.) You might even win a Nobel Prize.

As the eminent contemporary philosopher Philip Kitcher discusses in *The Advancement of Science*, scientists are often more motivated by achieving fame than by uncovering the truth. Moreover, this fact can actually make it more likely that scientists as a group will get to the truth. For instance, the American geneticist James Watson claims to have investigated the DNA hypothesis, not because he thought that it was true, but because it was more likely to make him famous if it turned out to be true. As it turned out, he got a Nobel Prize, and we got an amazing scientific breakthrough.

That is not to say that it is going to be easy to overturn the scientific consensus on global warming even if it is wrong. Moore probably overstates the case when he says that "any academic whose research dares question the 'settled science' of the climate change complex is instantly accused of being a shill for the oil and gas industry or the Koch brothers." But if you go against the received wisdom, the scientific community might think that you're nuts, and you could have trouble getting your research published. It may be that the truth will always prevail, but it can certainly be delayed. As renowned British philosopher John Stuart Mill (1806–1873) pointed out in *On*

Liberty, "history teems with instances of truth put down by persecution. If not suppressed forever, it may be thrown back for centuries."

But if anthropogenic climate change is false, there is good reason to think that compelling evidence for its falsity would have been found. Researchers who want to overturn the scientific consensus on global warming are well funded by the fossil fuel interests. For instance, Winsberg discusses how "the American Enterprise Institute offered a $10,000 prize for scientists who found evidence that contradicted" anthropogenic climate change. Also, researchers who have questioned the scientific consensus on global warming have gotten a fair hearing by climate scientists and not just by *Fox and Friends*.

The work of a group at the University of Alabama in Huntsville, which found no warming in the troposphere (the lowest layer of the Earth's atmosphere) based on satellite data and weather balloon data, has been published in the top academic science journals, such as *Science* and *Nature*. (Of course, as Winsberg describes, once the group's data was ultimately corrected based on our understanding of the measuring instruments, "the controversy was definitively settled in favor of" anthropogenic climate change.)

The End?

A vast global conspiracy of climate scientists is a bizarre idea. However, it's not the only way to explain why almost all climate scientists agree that humans are causing global warming even if we aren't. Maybe climate scientists are unduly influenced by the views of a few eminent scientists who happen to be mistaken. Or perhaps, climate scientists benefit financially from endorsing anthropogenic climate change in a way that unconsciously biases their research. But these explanations are pretty implausible too. So, I'd still bet that, unless we start taking these climate scientists much more seriously, we'll just have to raise a glass of Scandinavian wine to toast the end of the world.[1]

[1] I would like to thank Tony Doyle, Peter Lewis, Kay Mathiesen, and the students in my course on *Science and Pseudoscience* at Northeastern University for many helpful suggestions.

PART III

---

*"The best kept
secrets are those
in plain sight."*

11
Are Collectives More Conspiratorial than Individuals?

EDUARDO VICENTINI DE MEDEIROS AND
MARCO ANTONIO AZEVEDO

We'd bet good money that either you know someone who believes in conspiracy theories, or this someone is you. And it's okay.

We shouldn't make a fuss about it. Even if we were to consider the evidence that links belief in such theories with some kinds of mental disorders, cognitive biases, or unfavorable social conditions, there's no reason to believe that a person must be paranoid, irrational, or socially marginalized in order to believe in conspiracies. In some cases, we don't even suspect the hidden engines of conspiracies, and that unawareness is more likely a result of being politically unaware than of harboring a mental disorder. That's to say, we mostly don't know how power relations come about, and that is not because we are behaviorally, cognitively, or socially unable.

What Do We Mean by "Conspiracy Theory"?

Philosophers just love to start a conversation with an attempt to define concepts. So, let's delay this no further and pose the question: what are we talking about when we say, "conspiracy theories"?

Consider this proposal:

> Conspiracy theories are explanatory beliefs for events of great social impact that take them as effects of the activity of groups or organizations that, nevertheless, maintain their causal role hidden from the public opinion.

To test this proposal, let's see if it fits into some well-known cases. Let's start with the theory that the September 11th attack was an inside job perpetrated by the US government.

In this case, the "event of great social impact" is the fall of the Twin Towers. The organization involved in the production of this event is the US government or some of its organs and agencies, perhaps the White House itself, or, in other versions, the CIA. But, of course, no government agency would act against the lives and assets of its citizens openly. The way to do that would be to create a cover story that would pass through an official, which places responsibility with another "enemy" organization, in that case, al-Qaeda.

As another example, let's consider the assassination of President John Fitzgerald Kennedy. In this case, the event of great social impact is the assassination of a popular president of the United States. Against the official version (that the killer was a lone sniper), a conspiracy theory might claim that the responsibility lies with an organization, such as the KGB, or even the CIA, and is covered up by an official story created for public consumption: Lee Harvey Oswald acted on his own.

A third example is an invitation to descend a little on the map of the Americas, where the theory that "Lava-Jato" (Car Wash Operation, conducted by the Federal Prosecution Service of Brazil), which allegedly destabilized the political situation in that country, was in fact a hidden plan designed by the group known as the Seven Sisters Oil Company to take the newly discovered Pre-Salt Layers located in the Brazilian continental shelf. Here, there are several events of great social impact.

We'll list just two: the arrest of former President Lula and the privatization, authorized by the National Congress, of pre-salt layers exploration fields that were previously exclusive prerogatives of Petrobras, a state-owned company. The official story for the media, however, would be to take the Car-Wash Operation as an autonomous initiative of Brazilian judicial control bodies.

Let's now cross the Atlantic. The deaths of Princess Diana and Dodi Fayed were undoubtedly events of great social impact, and they provoked countless conspiracy hypotheses which contradicted official investigation reports pointing to a tragic, fatal, yet purely accidental car crash.

In some versions, the organization that planned the incident was the British intelligence service, the world-famous MI6, which would have acted in the interests of the British Royal Family, and which in turn would avoid the continuation of a supposed pregnancy of Diana—the fruit of her relationship with Dodi. Official versions maintain that the chauffeur, Henri Paul, was driving under the influence of alcohol and prescription drugs, and that the paparazzi also contributed to the accident. In place of those explanations that put individuals in charge, conspiracy theories suggested a collective arrangement, by the Royal Family moved by their private interest, as the best explanation for the fatality.

Why Do We Think that Collectives Promote Conspiracies?

We could increase the number of examples, and you can check for yourself the adequacy of the proposed definition using your preferred event with its respective conspiracy explanation.

For now, let's consider why it is that our proposed definition is appropriate. One of its advantages is that it marks a universal element in any conspiracy explanation: the effective participation of groups or organizations, not individuals, in the production of the relevant events. The relevant questions about the dramatic events above are: was it the CIA or the White House who planned the events of 9/11 rather than an individual, say, President George W. Bush? Was the CIA or the KGB behind the murder of JFK? Or possibly was it Lyndon B. Johnson or Nikita Khrushchev?

Has Seven Sisters Oil planned a far-reaching and intricate execution to promote the privatization of Brazil's natural wealth, and not one, two, or even seven of its CEOs? Was the death of Princess Diana a plot of Prince Charles, or of the British Royal Family, run by MI6? This raises the question: Why could the conspiracy activity not have been planned and executed by some psychopathic individual mind or a selfish authority alone?

Why don't people typically consider these possibilities? We think that people don't consider those possibilities for a very simple reason: it's cognitively less burdensome to create alternative conspiracy scenarios when what is at stake are the actions and intentions of a collective agent rather than an individual.

Individual agents are immersed in their biographies, filled with emotional idiosyncrasies, affective bonds, cultural restraints, family ties, and particularities of every order. Collective agents, on the other hand, have a publicly accessible institutional character, and their agencies are not filled with those personal contingencies. For those who accompany their public behavior, it is easier to grasp and think about their plans and intentions than to think about the intentions of individual persons with whom we are not usually acquainted.

Collectives are less receptive to the peculiarities of personal characters, and our bonds with collectives are significantly less personal than the bonds we have with the people we know that serve to us as models to imagine how others might behave.

Attributing Collective Intentions

Collective agents, including nations, associations, and groups, are not the mere sum of the individuals who compose them. They are not equivalent to the contingent collection of their members. After all, a collection can change its members without changing the collective it "represents."

We usually also attribute intentional actions to collectives. Nations can declare wars, crowds may engage in protests, parties can lead governments, and organizations can plan and execute terrorist attacks. But when we say that a nation declared war, we are not implying that all the individuals agreed and participated in the deed. Thucydides said that the Peloponnesian War began after a surprise attack of the Thebans against Plataea. Nevertheless, only a small group of individuals were involved in the act. See this passage from Thucydides famous book *The History of the Peloponnesian War*:

> On becoming aware of the presence of the Thebans within their gates, and of the sudden occupation of the town, the Plataeans concluded in their alarm that more had entered than was really the case, the night preventing their seeing them. They accordingly came to terms and, accepting the proposal, made not movement; especially as the Thebans offered none of them any violence. But somehow or other, during the negotiations, they discovered the scanty numbers of the Thebans, and decided that they could easily attack and overpower them; the mass of the Plataeans being averse to revolting from Athens. At all events they resolved to attempt it.

In Thucydides's description, The Thebans and the Plataeas are collective entities, while the "scanty" Thebans are individuals. Thucydides says that the Theban attack was carried out by this scarce number of individuals. But this fact does not lead us to conclude that it was not the Thebans who accomplished the action, for we ordinarily assume that collectives act by means of individual members. David Copp offers the concept of "secondary action" in order to explain that. An action performed by an entity—either an individual or a collective—is a *secondary action* if this action is correctly attributed to this entity on the basis of an action done by another agent or agents. Actions of collectives are always secondary actions, even though not all secondary actions are actions of collectives.

We defined a conspiracy theory as an explanatory belief for events of great social impact that attribute responsibility to organizations or groups that, nevertheless, keep their performance hidden from the public opinion. It is a characteristic of this kind of explanation that the agent must be a collective, and not an individual. But there is a reason why some people are prone to assign responsibility to collectives than to individuals in these circumstances. When we attribute an intention to a collective, we don't need to think about all of the personal characteristics of the individuals by means of which the collective is supposed to act. For example, when you get a traffic ticket you are not only fined by John or Peter but by the traffic police. Even if you know some personal things about John or Peter, their acts are not explainable by their characters. After all, they are traffic officers, and we understand their conduct from what we know about the way traffic officers ordinarily act, not from what we know about their particular lives.

So, consider the accident that killed Princess Diana. For those of us who are not acquainted with Prince Charles or Queen Elizabeth II, it is cognitively costly to imagine that Prince Charles had agreed to a plan to murder his children's mother (for we should know much more about him to consider that). Nevertheless, we don't need to strive cognitively too much to imagine that a collective entity, such as the Royal Family, could have planned that, since there are plausible reasons to believe that their members were immersed in the pomp and circumstance of an age-old tradition to be protected from scandals and unwanted family ties.

It is cognitively easier to entertain beliefs about the responsibility of collective entities than about individuals we are not acquainted with, since by assigning beliefs to collectives we do not commit ourselves to any knowledge about the particular minds of its eventual members. And we can keep a belief about the thoughts and intentions of collectives, even agreeing that some or a few of their members do not think in exactly the same way. We may even believe that individuals would be separately unable to undertake actions that, in a group, would nevertheless be inclined to perform, even with regret.

What we're suggesting is that the greater degree of mutability of our collective agency attributions can be an important explanatory factor for the construction, acceptance, and transmission of conspiracy theories. By *mutability* we mean this quality which makes the intentions attributed to an entity

more susceptible to being adapted than any preferred explanatory theory. Collective intentions are in this sense more muta-
ble than individual intentions.

Conspiracy Thinking and Academic Explanations

Different kinds of explanations have been proposed for the construction, acceptance, and transmission of conspiracy theories.
We can divide these explanations into three groups: *psychological*, *sociological,* and *epistemological.*

We have rejected the claim that beliefs in conspiracy theories
are only, or mainly, the result of some kind of psychological disorder or, in the limit, of mere irrationality. The conspiracy mindset
is not a direct consequence of any specific mental disturbance.
Indeed, in various circumstances, it clearly makes sense to consider the possibility that some conspiracy is going on. Anyone
who has watched *House of Cards* or who walks the backstage of
politics in the real world knows what we are talking about.

Watergate gives us an excellent example of an episode that
began with a conspiracy hypothesis that hypothesis ended up
providing an accurate description of the facts. Moreover, by
casually associating all or even most conspiracy explanations
with some kind of mental pathology, we also end up committing
ourselves to the assertion that an excessive number of people
in different corners of the world are indeed mentally ill, since
the numbers of people who adhere to conspiracy theories is
impressively large
<www.gallup.com/poll/165893/majority-believe-jfk-killed-
conspiracy.aspx>
<www.publicpolicypolling.com/polls/democrats-and-
republicansdiffer-on-conspiracy-theory-beliefs>.

However, if we want to explain why people believe in conspiracy theories, we cannot leave behavioral disorders and irrationality totally out of the equation. This is why we need the
help of social psychology. Karen M. Douglas and a team of collaborators have carried out detailed research on the subject
over the last few years, as well as useful bibliographical
reviews for a synoptic view of the topic
<www.kent.ac.uk/psychology/people/douglask>.

Some interesting research developed by Hannah Darwin,
Nick Neave, and Joni Holmes found positive correlations
between the belief in conspiracy theories and "paranoid
ideation" and "schizotypy," two embryonic forms of more severe
psychological dysfunctions.

Sociological explanations of different sorts have also been proposed. Karen Douglas and her collaborators reviewed this kind of published research in the years 2005–2017. Some experimental studies have suggested that experiences of ostracism cause people to believe in superstitions and conspiracy theories, apparently as part of an effort to make sense of their experience.

Other studies have shown that members of groups that experience low social status (because of their ethnicity or income) are more likely to endorse conspiracy theories. It seems also that people who suffer more political defeats than victories are more prone to believe in conspiracy theories. Conspiracy beliefs have also been linked to prejudice against powerful groups and those perceived as enemies.

Philosopher Quassim Cassam introduces the central notion of *epistemic vice*. People believe in conspiracy theories, says Cassam, because of the epistemic character they develop. The claim is that people who engage in conspiracy thinking exhibit certain behavioral dispositions, such as gullibility, wishful thinking, closed-mindedness, or prejudice. Hence, it's not only the quality of the information available for the epistemic agents that is defective, but also their "vicious" intellectual dispositions.

His account assumes that there must be epistemic virtues in addition to the vices. Nevertheless, the proneness to assign intentions to collectives rather than to individuals is not properly a vice in its own right. Epistemic vices are wrongful dispositions, yet there are many occasions in which we are not wrong to attribute intentions to collectives.

Secret Plans by Collective Agents

For the conspiratorial mindset, defining the real intentions of the collective agents or associations acting behind the scenes of power relations in such sensitive areas as economy, religion, or politics is always a stage for flights of the imagination. How many secret plans have been assigned to the Vatican, the Illuminati, the KGB or the CIA, the Seven Oil Sisters, the Bilderberg Club, or the Globalists? We want to draw attention to a common ground in these attributions.

In conspiracy theories, the main actors are mostly groups or organizations, that is, collective agents. Therefore, we would like to explore more deeper our hypothesis that it would be cognitively less burdensome to create alternative scenarios when what is at stake are the actions and intentions of a collective agent. Why is this cognitively less burdensome after all? What we're claiming is that the *mutability* of the intentional attribution to collective agency is more elastic in comparison with the intentional attributions to individual agents. But what do we mean by "mutability" in this context?

By *mutability* we mean a proneness to consider counterfactual (that is, alternative) scenarios relatively to collective agents compared with individual agents. This could explain why only counterfactual scenarios of collective agency arise in conspiracy theories against the background of what is often referred to as the "official account" of events of great social impact.

In order to make this hypothesis stronger we need to take a step back and consider some cognitive limits to what we can explanatorily imagine. The faculty of imagination, which we can understand as the ability to think about how things could have happened differently from how they appear to have actually happened, is subject to conditioning factors. These limitations were systematically investigated by Ruth M.J. Byrne, following the seminal steps of Daniel Kahneman, Amos Tversky, and Dale T. Miller.

The main question Byrne's research sought to answer is: Why do people imagine alternatives to some aspects of reality more readily than others? A good question, in the context of a discussion about production, acceptance, and transmission of conspiracy theories, can be formulated in similar terms: Why do people imagine alternatives to the intentions of collective agents more easily than alternatives to the intentions of individual agents?

For now, we only consider one possible answer to that: people's proneness to explain events of social impact in a way that indicates that some collective is hiding their responsibility from the public can be partially explained by the fact that it is easier to attribute intentions to collectives than to attribute them to individuals with whom we are not acquainted.

This is at most a partial explanation; so, we do not have any complete answers to those questions. As soon as empirical research can offer us better answers, it may become clearer why conspiracy-minded people are so prone to those explanations even against the available evidence, such as: what if the explanations offered by the media about the collapse of the Twin Towers were false? What if the Seven Oil Sisters had devised a plan to withdraw the Pre-Salt reserves from Brazil? What if the official story of Princess Diana's death was only told to cover up MI6 actions at the behest of the Royal Family?

What if Lee Oswald was a KGB undercover agent working to undermine democracy in America?

In his classic *The Wealth of Nations*, Adam Smith famously warned:

> People of the same trade seldom meet together, even for merriment and diversion, but the conversation ends in a conspiracy against the public, or in some contrivance to raise prices.

Would Bilderberg Club meetings provide just one more example of the phenomenon described by Smith?

12

Here's the Secret on Voter Fraud—It's Complicated

GARY JOHNSON

Allegations of election-related fraud make for enticing press, and fuel many Americans fears that elections are "rigged." Many Americans remember vivid stories of voting improprieties in Chicagoland, or the suspiciously sudden appearance of LBJ's alphabetized ballot box in Texas, or Governor Earl Long's quip: "When I die, I want to be buried in Louisiana, so I can stay active in politics."

Voter fraud, in particular, has the feel of a bank heist: roundly condemned but technically fascinating, and sufficiently lurid to grab and hold headlines. Perhaps because these stories are dramatic, voter fraud makes a popular scapegoat and enduring source of conspiracy theories. In the aftermath of a close election, losing candidates are often quick to blame voter fraud for the results. Legislators cite voter fraud as justification for various new restrictions on the exercise of the franchise. And pundits trot out the same few anecdotes time and again as proof that a wave of fraud is imminent.

Allegations of widespread voter fraud, however, often prove greatly exaggerated. It is easy to grab headlines with a lurid claim ("Tens of thousands may be voting illegally!"); the followup—when any exists—is not usually deemed newsworthy. Yet on closer examination, many of the claims of voter fraud amount to a great deal of smoke without much fire. The evidence simply isn't there. The US Election Assistance Commission, the Department of Justice, the Heritage Foundation, and the Brennan Center for Justice all come to the conclusion that voter fraud is extremely rare.

On the other hand, voter fraud does exist, not just in America, but also in just about every election since the beginning of time (to be fair, this is a logical assumption and not a fact) including student council elections. Make no mistake, neither the left nor right (radicals excluded) think we don't need to protect our democracy from voter fraud, they just have very different opinions on how this should be done. The Texas Secretary of State's office is quietly telling counties that many of the voters it flagged for a citizenship review are actually citizens, after earlier declaring publicly that 95,000 voters were illegally registered to vote, according to the *Texas Tribune*.

In truth, as of this writing, there are four demonstrated examples of people committing voter fraud during the 2016 general election (none of whom were prosecuted) consistent with the last few national election cycles where instances of fraud are often one voter testing the limits of voter security. These examples include a woman in Iowa voting twice to prove the polls were rigged, a man in Idaho who filled out an absentee ballot for his dead wife, and a man in South Carolina who voted both absentee and in person. That's 0.000002 percent of the ballots cast in the race for the White House in 2016—if they counted, which they didn't. Looking back to 2000–2010, of the 649 million total votes cast there were thirteen instances of suspected voter fraud (during the same time period there were 47,000 reported UFO sightings in the United States, and 441 deaths by lightning strikes).

But the facts don't seem to matter. On Sunday January 27th 2019, President Trump declared via twitter that "58,000 noncitizens voted in Texas" and claimed that "voter fraud is rampant," a long-standing Republican claim that lacked any proof. Texas Attorney General Ken Paxton also tweeted the news with "VOTER FRAUD ALERT" in capital letters. This is nothing new.

Conspiracy theories influence debates over many policies; vaccine regulation, food labeling, campaign finance reform, immigration—you name it. But elections are different, more important, sacrosanct even. Elections are constitutionally required on a regular basis, with the "time, place, and manner" to be determined by state legislatures. In America, state legislatures administer all elections. Free and open elections are assumed by most voters, especially since the Civil War, most certainly since the Voting Rights Act of 1965, the Twentieth Amendment, the Seventeenth Amendment—I could go on but you get the point: we've worked hard for free and fair elections and we've got the scars to prove it.

Now, as to free and fair elections where all votes are counted and counted accurately, the political science literature would say we still have a long way to go. No institution, or practice, whatever you want to call it, is more vital to American Democracy than elections. No process is more subject to the belief that shadowy figures behind the scenes manipulate the results in their own favor. However, election processes are inherently subject to errors and are historically subject to manipulation and fraud. These processes therefore require extraordinary integrity (especially for any computerized systems involved), as well as honesty and experience among people involved in administering elections.

People often speak of the "right to vote" but there are actually two recognized rights. The first is the electoral franchise, the right to cast a ballot that is counted to determine the winner, and the second is the right not to have your vote nullified by an illegal vote. In fact the Supreme Court has recognized that "the right to vote is the right to participate in the electoral process that is necessarily structured to maintain the integrity of the democratic system" (*Burdick v. Takushi*, 504, U.S. 428, 441 [1992]). These two recognized voting rights are often in tension but the technology now exists that if adopted by all states could defuse voting conspiracy theories.

We could confirm voting eligibility and legitimacy in every precinct in America if there was a national law requiring a biometric identification card with picture, address, fingerprint or retinal scan that would authenticate every voter and cross reference them on a national database that integrates national, state, and local voter rolls. Anything short of that and there will always be questions and conspiracy theories regarding voting fraud. Such a system would of course raise serious privacy issues and be hugely expensive for the county governments that administer elections, the level of government with the fewest resources.

HAVA Heart

Those who cast the votes decide nothing. Those who count the votes decide everything.

—attributed to JOSEF STALIN

Because of the 2000 presidential election controversy, the subsequent Supreme Court's *Bush v. Gore* decision and similar voting problems in the 2002 midterm elections, Congress in December 2002 passed the Help Americans Vote Act (HAVA). This law requires states to regulate and enforce a variety of new regulations, minimum voting standards, and new federal programs. Among its sweeping provisions HAVA requires all states to provide voters with opportunities to cast provisional ballots; access for persons with disabilities to the voting place, voting information, such as sample ballots, voting instructions, and a statement of voters' rights; opportunities for voters to verify their selection, correct any errors, and notice if they overvote; and procedures for voters to make complaints when voting problems arise. The law also compels states to require identification and to verify residency requirements for new voters; to create statewide computerized voter lists; to eliminate punch card voting systems (the most common voting machine in America at the time); to train poll workers in the law's new requirements (in the 2016 election 24 percent of poll-workers were over age 71, another 32 percent were aged 61–70) and include check-off boxes for US citizenship and being eighteen years of age on all mail-in voter forms. HAVA allocated $3.86 billion to fund new equipment, advance accessibility and improve administration—however HAVA did not require a uniform voting system (which most other Western countries have), but does set mandates about which system local officials choose. For federal elections HAVA requires error correction for voters, manual auditing, accessibility, alternative languages, and federal error rate standards.

Many requirements of HAVA for subnational governments are not funded. That is, state governments must find room in already limited budgets to enforce these measures. Thus, they usually pass these substantial expenses on to local county governments where elections actually take place. Counties are the least equipped to absorb these expenses and have little incentive to pay for expensive technology and infrastructure improvements that get used once every year or two. HAVA partially funds the creation of a new federal Election Assistance Commission to study problems and make recommendations for improvement in the way America votes, and establishes a federal program to expand poll worker recruitment, providing federal money for states to implement HAVA's provisions and to improve the administration and reliability of elections. HAVA also requires states to have replaced punch card and lever operated voting machines before the 2004 general election. There is little agreement on which technology is best, or whether HAVA "worked" in the 2018 midterm elections. The general consensus is that HAVA has done little to substantively improve the accuracy and convenience of American elections, although there is evidence that it helps newer and poorer counties with their accuracy.

Our election system is unlike any other democratic society and is bizarrely complicated. Each of the fifty states has a different system for electing officials. Some states require voters to be registered months in advance before election day, whereas others (such as Wisconsin) allow voters to register when they show up at the ballot box. North Dakota has no requirement at all for registration. Some states mandate that voters register as members of a political party and others do not. Some states require political parties to have open primary elections for selecting party candidates, whereas others allow party conventions or caucuses to accomplish this. And as California demonstrated in 2003, some states have provisions for recall elections.

Since 2004, the Election Assistance Commission has conducted the Election Administration and Voting Survey that asks all fifty states, the District of Columbia, and four US territories—American Samoa, Guam, Puerto Rico and the Virgin Islands—to provide data regarding the way Americans vote in each federal election. This is a data goldmine on election administration in America. It provides feedback and a consistent reliable data source for policymakers, election administrators, and the general public. In the 2016 election, only thirty of the 6,467 jurisdictions in the survey did not respond to the survey. We therefore have reliable data sources, and do not need to speculate on voter fraud. Of the 140,114,502 citizens who voted in the 2016 general election we had a turnout rate of 63 percent of the citizen voting age population.

How We Vote

The technology used for casting a vote varies tremendously. Some states, and counties within states show even more variation, using paper ballots. Others use levered voting machines. Still others use high-tech touch screen computers. Regardless of the technology currently in use HAVA requires all counties to provide voters with a paper receipt. So, as a result of state control and management of elections, fifty unique systems of voting can be found across the fifty states. And, when we consider that there are 3,142 counties in the United States (as of April 2019) this configuration makes it difficult to summarize and explain how votes are counted—however it does provide for a great deal of color, nuance, and diversity in American campaigns and elections. The problem is that in most counties we have no clear standard put in place to ensure that votes are counted, and counted accurately, and there are serious questions regarding the security and accuracy of all available electronic voting machines.

The Nature of Conspiracy Theories

Richard Hofstadter (1966) noted that a core function of a conspiracy theory is to provide straightforward simple explanations for complex and distressing events that are hard to comprehend otherwise. There is ample evidence in the political science, philosophy, and psychology literature that increased levels of education decrease an individuals' tendency to accept conspiracy theories, and hopefully I've made the case that voting is quite complex and confusing in America. A study by Van Prooijen et. al (2015) found conspiracy theories are strongly associated with simple solutions for complex social problems and that formal education reduced this effect.

When it comes to politics, conspiracy theories need to be put into a broader context than the "paranoid style of American politics" so prone to outrage, distortion, exaggeration, and outright nonsense. We have to recognize that there have been widespread conspiracies in American history and that the cultural (movies and television in particular), institutional, and technological changes since Watergate and the great paradigmatic shift it portended in the way Americans view government did not take place in a vacuum and have made conspiratorial thinking a mainstream phenomenon.

During World War I according to Kathryn Olmsted conspiracy theory underwent a fundamental transformation. Lies, cover-ups, illegal surveillance, assassination, political "dirty tricks" became the official means to control events and media interpretations of events. Olmsted thoroughly documents the twentieth-century rogues gallery of federal chicanery: misinformation campaigns by Wilson and FDR, J. Edgar Hoover's anti-communism and the FBI's COINTELPRO's work to silence and spy on perceived dissidents, JFK's assassination attempts on Castro, Watergate, the CIA's MKULTRA plot to use LSD on unknowing subjects, and the Bush administration's conspiracy to lever the 9/11 attacks into a war with Iraq (this is a very brief list). In short, just because you're paranoid doesn't mean they aren't out to get you. It is increasingly hard to defend democratic institutions, like elections, when so much of our perception of the activities and motives of the federal government are toxic.

The conspiracy theorists' fire is not without fuel— Americans have good reason to distrust federal government and its institutions, and yet most Americans who are educated and law-abiding believe in the integrity of our elections. The federal government can always improve its transparency, accountability, and oversight functions to make government trustworthy, it should be said. Yet in an age where we see conspiracy everywhere and have it presented in the media and culture as the standard, not the exception, it's no wonder voter fraud conspiracies endure. Conspiracy theories are by definition false; instead, many real conspiracies have come to light over the years and for many Americans seem to be the way government does business. Suspicions of President Nixon's involvement in a burglary at the Democratic Party National Committee headquarters were dismissed as outlandish conspiracy theories at the time by the bulk of the media and as measured by public opinion polls, but turned out to be true. Even when proven wrong however, conspiracy theories are notoriously resistant to falsification—there's always room for their proponents to claim that the new layers of the conspiracy have not yet been revealed (the "it's even bigger than we thought" effect). The psychology literature on cognition is illuminating. Their excellent research on belief in conspiracy theories concludes that belief in a particular theory is strongly correlated with a belief in other conspiracy theories, leading to the finding that belief in one conspiracy is a predictor of other conspiracy beliefs that come together in a mutually supported network labeled a monological belief system. Voter fraud fits nicely into this worldview of a government ruled by a massive sinister "deep state" that controls all major institutions in our society. For a political scientist like me this is easily falsifiable and counterintuitive. Our system is so fragmented and has so many access points that such coordinated deception seems impossible.

Scholars who study elections tend to focus on two key values to measure the legitimacy of the process—integrity and access. Given that states and local governments vary so widely in how they do the incredibly complex job of running elections it's no wonder conspiracy theories about voter fraud abound, especially on the Internet where like-minded individuals can stoke each other without encountering alternative views. There are around ten thousand sub-state jurisdictions in the United States who are responsible for qualifying candidates and issues for ballots, designing ballots, training poll workers, keeping voting rolls up-to-date, and validating election results—all within a context of tight deadlines in a rigid legal framework and scarce funding.

For decades, political scientists have written about the "rational voter," a hypothetical person who takes in information about a candidate, parties, platforms, and policies and supports the ones that will improve their lives, usually in a material sense. The more threatening and alarming charge against the way we vote is not voter fraud, but, rather, the more recent research in political science suggesting that voters either don't have enough information to make rational decisions, or optimal policy choices, and instead vote according to group identities and social attachments (Achen and Bartels, *Democracy for Realists*).

The major parties are able to tap into ever more powerful databases to target their supporters and tailor their campaigns and policies meant to appeal to them, giving huge advantages in ballot access. The advantage is stability—frequent elections create stability and predictability because even losers have another chance next year to promote and defend their candidates and policies. However, if we look at voters' actual information and actual motivations, we are a long way from the fabled "rational voter."

13

Vaccination, Autism, and Trust

KENDAL BEAZER

In 1796, Edward Jenner successfully tested his theory that cowpox gave an increased immunity against the more virulent smallpox virus. The strain of cowpox that would be given to people around the world was termed "vaccinia," and this was the birth of vaccination.

It led to the eradication of smallpox and the near-eradication of polio. Diphtheria, tetanus, measles, mumps, rubella, and *Haemophilus influenza* type B have all been reduced by more than ninety-five percent of the morbidity that was seen prior to the twentieth century. This tremendous success of modern medicine has recently experienced a reversal due to a small, yet vocal anti-vaccination movement. This group argues that the side effects of vaccinations are being hidden from the general public. The anti-vaccination community has developed a large distrust towards the healthcare community and believes that the government has covered up data that links autism spectrum disorder (ASD) to the measles, mumps and rubella vaccine (MMR). Underlying this theory are multiple levels of trust and distrust that are deep-seated in our history.

History of Vaccination

Edward Jenner was an English physician who had grown up in a farming community and knew the rumors that milkmaids did not contract smallpox. In their work with cattle, they would often contract the much milder disease, cowpox. Jenner hypothesized that a bout of cowpox would make a person immune to smallpox.

In order to prove his theory, Jenner removed pus from cow-pox lesions of a milkmaid by the name of Sarah Nelmes. Jenner then took this pus and injected it into the cut of James Phipps, the son of a poor laborer. James spiked a fever, but soon after became well again. Several days later, Jenner injected James with smallpox and James remained healthy. Jenner named the method vaccination, after the word vacca, Latin for 'cow'. Over the next couple of years, Jenner repeated this vaccination method with several other people, including his own son. His method was met with much skepticism and it was not until 1853 that vaccinia became routine treatment for smallpox in England <https://www.historyofvaccines.org/timeline/all>.

While he knew that vaccination worked, Jenner had no explanation for how it worked. At this time miasma, or dirty air, was the primary explanation for illness. While there had been a lot of work done to disprove miasma theory and prove the germ theory of disease, scientists remained skeptical. It was not until the epidemiological work of John Snow with the Broad Street Pump, Louis Pasteur's fermentation experiments, and Koch's postulates, that germ theory became recognized as the primary explanation for communicable diseases. Since then, significant amounts of time and money have been invested in stopping the germs that are behind communicable illnesses. The 1930s through the 1960s saw the development of numerous vaccines. Combine vaccination with the discovery of antimicrobial drugs in the 1930s and 1940s and it was a remarkable victory in the ongoing battle against microbes.

Prior to these advances in the knowledge of microbiology, death from a variety of different diseases were common. With a thirty percent mortality rate due to smallpox, it has been estimated that five million lives per year have been saved since 1980 due to this vaccination alone <https://ourworldindata .org/smallpox>. Diseases such as diphtheria, tetanus, pertussis, haemophilus influenza, measles, and streptococcus pneumoniae had between 0.1 and 13 percent mortality rates. These rates do not take into account lifelong illness, or secondary diseases that may accompany the initial infection. Vaccination drastically changed morbidity and mortality rates and is one of the most successful public health initiatives ever to be implemented.

MMR Vaccination

Prior to the measles vaccine, nearly all children came down with the disease before they were fifteen years of age. This was a nationally notifiable disease which required every healthcare facility to report each case they came across and it is estimated that three to four million people were infected each year. Among the reported cases, each year 400 to 500 people died, 48,000 were hospitalized, and 1,000 suffered encephalitis (swelling of the brain) from measles. Due to vaccination, reported measles cases were reduced by 80 percent in 1981, and the disease was declared eliminated from the United States in 2000 <www.cdc.gov/measles/vaccination .html>.

Before the mumps vaccine was routinely used, around 186,000 cases were reported each year. This disease is rarely fatal but causes fatigue, headaches, and a low-grade fever that lasts a few days. Other complications that can arise are deafness and swelling of the brain. Though we still see the occasional outbreak among vaccinated people, the incidence of this disease has decreased by 99 percent since the introduction of a vaccine for mumps. <www.cdc.gov/mumps/vaccination .html>.

Rubella is a very contagious disease that comes with a mild illness. This includes a low-grade fever, sore throat, and a rash that starts on the face and spreads to the rest of the body. Similar to mumps, rubella is not associated with a high mortality rate, but is very dangerous for a fetus. Rubella can cause miscarriage or birth defects in an unborn child. The last major outbreak of rubella in the United States happened in 1965 and resulted in 12.5 million cases. Due to vaccination, rubella was declared eliminated from the United States in 2004 <www.cdc.gov/rubella/vaccination.html>.

Of all vaccines that are used today, none has caused more controversy than MMR. This vaccine was licensed and put into use by the United States government in 1971. MMR was a highly successful vaccine that induced immunity to measles in 96 percent of vaccinated children, rubella in 94 percent, and mumps in 95 percent. Even with the success of vaccination, suspicion has been around since its discovery.

In 1926, mandatory vaccinations had become very unpopular in Georgetown, Delaware. A group of health officers visited the city to vaccinate the townspeople; a retired army lieutenant and a city councilman led an armed mob to force the health officers out of the city, effectively stopping their vaccination efforts. This goes to show that vaccination was unpopular long before the MMR and autism controversy began. However, the events that unfolded at the end of the twentieth century were the most detrimental to the vaccination movement.

The Wakefield Study

In 1998 a British physician, Andrew Wakefield, and twelve colleagues published an article in the journal, *The Lancet*. This article suggested a link between MMR and autism. Even though the study was flawed from the beginning with a small sample size, poor design, and a speculative conclusion, worry spread like wildfire. The seed of distrust had been sowed. His explanation gave hope, and a target, to a group of people that felt like they had lost their children to ASD. Andrew Wakefield, being a physician, was naturally in a position of trust and he published his paper with twelve other physicians and experts who also garnered trust on this issue. Their newly found connection was devastating to MMR vaccination and eradication efforts.

What came in the following years was an enormous amount of research. The MMR immunization was analyzed multiple ways. Researchers attempted to replicate the Wakefield study, but no correlation between MMR and ASD was found. In contrast, one of the largest studies was done in 2002 on ASD and MMR. This was a retrospective cohort study for all children born in Denmark from 1991 to 1998. It involved looking at the medical history of 537,303 children. The results of the study are as follows: "There was no association between the age at the time of vaccination, the time since vaccination, or the date of vaccination and the development of autistic disorder" (*New England Journal of Medicine*). Even aside from the superiority of the research design, the sample size was enough to overshadow Wakefield's studies.

As research continued to be published refuting Wakefield's assertions, scrutiny fell upon the man behind the paper. Ten of the twelve original authors retracted their interpretations from the original Wakefield paper and after ten years, his study was revealed to be completely fraudulent. In 2010, *The Lancet* retracted the article on the grounds of ethical violations and scientific misrepresentation. Shortly after the retraction issued by *The Lancet*, Andrew Wakefield was removed from the United Kingdom medical register, effectively ending his career as a physician (Rao and Andrade).

Wakefield has continued to advocate for what he believes and has continued to pick up supporters along the way. In 2016, his company released the movie *Vaxxed* that alleges a cover-up of the MMR-ASD link by the Centers for Disease Control and Prevention. Wakefield's work over the years has been bolstered by statements from celebrities and politicians, who themselves lack any scientific training, but have enough followers that are influenced by their opinions. Jenny McCarthy, Rob Schneider, Jim Carrey, Robert F. Kennedy Jr., President Donald J. Trump, and many other prominent celebrities and politicians have all proclaimed connections between autism and vaccinations.

Mounting evidence contradicting the autism and vaccination link has not deterred many anti-vaccination groups, and their successes can be measured through outbreaks seen in under-vaccinated communities. Measles outbreaks in Europe, Japan, Philippines, and the United States have all been linked to under-vaccinated populations. It has become such a growing problem that the World Health Organization has listed vaccine hesitancy as one of their top ten global threats this year. Smallpox is the only disease that we have been able to successfully eradicate, and it was all due to the invention of vaccination. We could eliminate several more diseases, such as guinea worm, polio, elephantiasis, measles, mumps, and rubella. These diseases could be eradicated, but this is not likely until MMR vaccination rates improve <www.cdc .gov/mumps/vaccination.html>.

Trust

The propensity toward trust is something that is little thought about in day-to-day life but is also necessary for a properly functioning society. Without the ability to trust a neighbor, we would live in continual fear of what could become of our home or our personal safety. Trust permits us the time and energy to devote to other areas of our lives and is an evolutionary tool for the advancement of a species. There are many levels and varieties of trust that hinge upon the relationship of the trustor to the trustee and the value that is being held within that trust.

The more value that we assign to a given possession or relationship the more vulnerable we will be, and the more trust that will be required to ensure its safety. I would leave my drink and pizza at a food court table to go to the restroom, knowing that my pizza could be stolen, but I would not leave my child. To this end, trust has a limit. Strangers could exploit this vulnerability, but the degree of vulnerability is the measure of trust that we have.

From a healthcare standpoint, trust is crucial to the eradication of diseases. It's not possible for every single person to be involved in treating patients with diseases or with the development of vaccines. It would be great for everyone to be able to see why it's necessary. It would alleviate distrust if everybody could sift through data and see what progress has been made, and how far we must go to remove these diseases from the planet. The education required to perform in these areas is prohibitive to that being possible. It must be simplified down to a level that does not require professional knowledge to process. So, what the World Health Organization and the Centers for Disease Control are saying is, "Trust us." The implication of a "Trust us" approach addresses the problem of the relationship between parents and the government. This is an invitation that cannot be taken at will or cannot be accepted voluntarily. Either we do trust the government, in which case it only serves as a reassurance, or we do not, and it only increases the divide (Annette Baier, *Trust and Antitrust*).

Trust is the reliance on another's goodwill. When we trust somebody, we leave open the opportunity that they will betray that trust. Trust may be required on behalf of one party but does not require it to be reciprocated by the other party. To trust somebody does require an expectation of goodwill or at least a lack of ill-will towards the individual. A simplistic explanation of trust in this case is an accepted vulnerability toward another individual's possible, though unexpected, ill-will. What makes a person trust or not trust is varied levels of this vulnerability, and the confidence we have in the goodwill of another.

It's human nature to place value upon the people and things around us and inherent in that value is our vulnerability. We tend to place a higher value on people and things that cannot be easily replaced or sustained on our own. Things that fall into this category would be our health, reputation, and our offspring. For example, you would likely let a medical intern borrow your calculator but would be far more hesitant to let them operate on you. The damage that can be done is not worth the vulnerability.

Why trust and be vulnerable with what we care most about in this world? It is out of necessity that we make ourselves vulnerable to others. Since the beginning, mankind has recognized the value in survival as an endeavor that is most successful when others are trusted to take responsibility in different roles within that community. The same is true today. No person could realistically survive as easily on their own as they could within a community. To believe that would require one to have a baseline knowledge of all necessities of life. To survive, they would need to know agriculture, hunting, construction, and preservation of health. Healthcare is of utmost importance because the need for immediate care in times of crisis give rise to moments that expose vulnerability and capitalize on trust.

The level of trust between a parent and a child is the most sacred of all in this vaccination conundrum. Societal conveniences make it much easier to raise a child than it would be to raise them alone. We trust educators within a school system to teach them, police and firefighters to keep them safe, markets and restaurants to provide nutrition, physicians and dentists to take care of their health, and so on. Many of the people that we rely on are governmental employees, and while some of the services provided are a convenience rather than entrusted, we still place value on the services.

The issue with trusting in the government is what to entrust them with. Inherent in vulnerability is the balance of relative power between the trustor and the trustee. Vulnerability and the relative power that is exhibited is determined by the level of dependence we have on that service. Which services are worthy of vulnerability? We may be capable of driving in the snow, so not having a snow plow clear the road does not leave us vulnerable. This shifts the relative power that the Department of Transportation has over an individual. Control over the situation decreases the vulnerability and relative power. Decreasing these then decreases the necessity of trust and the possibility of distrust in that department.

Trust is difficult to gain and maintain, but it is very simple to lose. The government of the United States has shown at times that it is not worthy of trust. Just a few examples are the Tuskegee experiments, the handling of Native Americans, and slavery. The Tuskegee experiment was government backed and was intended to study untreated syphilis cases in African American men. The participants were misled and did not understand what they had agreed to. After the study started, penicillin was discovered to be highly effective for treatment of the disease and was withheld from the participants for more than twenty years. The study was stopped when journalists published the story and created a public outcry. So, it is understandable that people may be skeptical of the good will that is offered by the US government.

Relative Power and Trust

Is a vaccination mandate the best approach to eradication of diseases? From a trust standpoint it further erodes a trust relationship due to the power dynamic being held over the parent's head. Vaccinating children should be considered a common good, where both parties have an interest in promoting the well-being of the child. Mandates shift the trust dynamic. The government and parents both have a common goal: the health and well-being of children. However, the dynamic changes because the government is looking after the collective population instead of each individual child, or that one parent's child. If the parent comes to believe that their child's health is a gamble of the collective population, then trust cannot be achieved. Suspicion of the trusted will take on operative motives that conflict with the trustworthiness of the government. This is exacerbated by the power dynamic. Instead of a common good, society is left with a strong power that threatens to cause harm, though minimal, to the trustor for lack of compliance. When this happens, trust is gone.

Trust among mentally competent adults is far different from the dynamic of trust among parents and their children. The trust that infants place in their parents can be compared to the trust in God. Religious people may consider themselves at the will of a higher power with unwavering trust. Whether good or bad, this higher power still loves the trustor and so long as that perception is maintained then trust will thrive.

As children grow, they see their parents have weaknesses and that they have the ability to make their own decisions, this diminishes the level of trust and takes on a new form in adolescence. Trust in the government can be compared to the adolescent age of trust in a parent. At this age, there is more skepticism of the good will that is being shown by the parent. Are the motives of the parent really in the best interest for the child or the parent? Are vaccinations really in the best interest of the child or are they in the best interest of society as a whole? Are parents to believe that the government cares about their individual child more than the population? It is illogical to believe that, so it pits the strong versus the weak in a power struggle over individual rights.

Trust and Social Media

Over the last twenty years there has been a shift in the power dynamic due to social media. A single individual who would have been relegated to shouting their beliefs on a street corner, has now been given an amplified digital voice to reach millions of people. Distrust in vaccination has found a strong foothold and been able to thrive here. The vaccination movement has found resistance in the form of Internet memes that are shared among friends, family, and those of similar interests or beliefs.

These memes are not fact-checked, or regulated in any way to prove their validity—not just by the anti-vaxxer movement, but also the pro-vaccine crowds.

It is a coercion match to compete for those without strong opinions by the means of witty, funny, or emotionally charged snippets of information. It is fascinating to look at trust as it exists in social media. Why are people willing to trust in the information that they come across in social media more than they are willing to trust in the information obtained from medical personnel?

Power and anonymity have allowed the dispersion of false information to thrive on social media platforms. Power is shifted in the online environment. You do not have to be an expert or have a lot of experience to be able to share your opinion, what matters most is the amount of people who are willing to listen to you. The number of followers you can acquire can be seen as a trust relationship. You will follow them if they continue to deliver content that supports a mutual belief that is held. The good that is being entrusted in this relationship is our reputation. Most people do not want to come across as moronic, but many of the memes that are shared by them are unbelievable when observed through a non-partisan lens.

Anonymity takes away the risk of distrust by allowing a person's reputation to not be harmed by their online accounts. Online accounts can be closed and re-opened under another name, and as such, lack permanence and accountability. This online community allows you to find like-minded individuals and will even suggest more accounts that are similar to the ones you have followed. This enables a person to become completely entrenched in the MMR-ASD conspiracy theory. It no longer seems like a crazy idea, if everyone around you believes it too. Trust in the content on social media is of low risk to betrayal. It validates beliefs in the face of contradicting facts and can allow one to grow their online reputation and self-worth.

One last point that needs to be addressed is the immediacy of illness and vaccinations and the strain that puts on trust. When you are in the midst of a measles outbreak and you are seeing friends and family being hospitalized and dying from complications of the disease, feelings of helplessness would be more prevalent. Upon the introduction of the MMR vaccine, these diseases essentially disappeared. We are in an era where people of child-bearing age have no recollection of what it is like to live in an era when these diseases were endemic. Parents in society today are being asked to trust the government and vaccinate their children for these diseases that are not high on their list of concerns.

Trust is a valuable concept in society. It is the platform upon which we can achieve amazing things, but it can be fragile. Any hint of dishonesty or hidden truths and it will perpetuate a distrust that has the means of rotting the platform for growth. Mankind teeters on the cusp of eliminating many diseases, but struggles to move over the hump due to a lack of trust in the government. Perhaps addressing the trust dynamic between a parent and child is more useful than addressing the societal effects as a whole.

We need to better explain how this vaccination will specifically benefit the parent and the child. Pediatricians still carry a large amount of trust and the parent's concerns should be addressed by them. It is easy to pitch all of the positives of vaccination, but in the current battle over vaccinations, the knockout punch of positive statistics without any acknowledgment of side effects will create issues with trust. Skeptics will believe that the government is hiding something. Trust plays a crucial role in our attempts to eradicate diseases, and without it, many of these diseases will persist in our communities despite our best efforts to eliminate them.

14

Leninism, Astroturfing, and Conservatism

Daniel Krasner

By the end of the nineteenth century, Marxism was facing a crisis. The roots of that crisis lay in the legacy of Marx's teacher Hegel. Marx had taken from Hegel his view that world history had a purpose of its own and that purpose would be achieved by transcending "contradictions."

Though Hegel sometimes used the word "contradiction" literally, when he wrote about about history "contradiction" meant conflict and suffering. The world's full consciousness of contradiction took the form of violent ideological conflict, which would inevitably result in a new form of consciousness containing the best of both contradictories. And, if it now seemed impossible that anything could contain the best of both democracy and non-democracy since they are mutually exclusive, that just showed it was impossible to understand a new form of consciousness until you had transcended the old. But this new paradigm had to be an improvement, because it ended suffering and conflict, which people desired.

History was thus a long advance from the absurd into the unimaginable. Hegel seems to have been abusing the logical principle of *ex contradictione* (or *ex absurdo*) *quodlibet sequitur*, which William of Soissons advocated in the ninth century. Literally, it means "Whatever you want follows from a contradiction (or absurdity.)" More loosely, it says that you can logically derive any sentence whatsoever from a sentence like "it's democratic and not democratic."

Logicians and laymen generally take it as a reason to avoid contradicting themselves. Hegel and Marx seem to have taken it as a reason to make everything contradictory—after all, weren't they promised "whatever you want"? (The catch, of course, was that anyone who embraced contradictions whole-heartedly became so incoherent that he or she couldn't figure out what he or she wanted.)

Marx also agreed with Hegel that this process, after a final period of suffering and violent confrontation, would end in a world with no "contradictions." Marx added that the relevant conflicts would be economic (Hegel preferred military ones,) and that the climactic bloody struggle was imminent, and the final utopia near. History and the proletariat, however, were not following the script Marx had written for them. Human suffering and class conflict were not increasing, and capitalism was not approaching a crisis. That may be why Marxism had been generally abandoned, even by originally Marxist parties, throughout the West, except in Russia, its most backwards country.

Lenin had a simple, audacious solution to this. If the proletariat—the class of wage workers—wouldn't do what it was supposed to, he would have it done for them. He would take the most "class conscious" proletarians into his Bolshevik party (where Lenin decided who the most "class conscious" were), organize a militia from them, and use that militia to effect his revolution. This would be even better than relying on the proletariat to do it. The class-consciousness of most Russian proletarians was just too low, and they would only make a mess of things. Lenin's Red Guard had the same class-consciousness as they did, only higher, and, so, was entitled to think for them (since they couldn't think for themselves) act for them (since they couldn't act for themselves) and martyr them (since they couldn't martyr themselves.) All proletarians were equally proletarian, but Lenin's Vanguard were more equal than others.

And so Lenin instigated a putsch by a small, but highly trained and disciplined paramilitary force, against a weak, but generally accepted, national government, and sold it to the world as a mass movement. As far as I know, it was the first important example of faking a grassroots movement, a practice that was named "astroturfing" many decades later.

This bloodstained fraud was the Original Sin of Leninism. Leninists have taken it as the source of their authority, putting them in charge of all historical developments since, and allowing them to astroturf all grass roots movements since.

Lenin originally intended his astroturfing to be a one-time thing. There was a little hitch in Marx's grand plan, but, once they'd gotten over this bump in the road, Objective Historical Forces would kick in, the Proletarian Revolution would spread to the rest of the world, and all political and theoretical problems would be solved, just as Marx had said. Instead, the Russian Army dissolved, the German Army strolled through Russia unopposed, Russians splintered into countless warring factions, and even the war-weary West found the spirit to invade sometimes. The international proletariat was unwilling to follow their example.

In response, Lenin did a number of good things that would have been better if he'd managed to admit he had been wrong. He created a Soviet military, reunited most of the Russian Empire, gave his state a strong central government, and allowed some capitalism to exist inside it under his New Economic Policy. What he did not do was repudiate the boastful promises he'd made for his "Revolution." If the Bolshevik Revolution was merely a bloodstained fraud, then the members of the Party were merely bloodstained frauds, and he and they would have lacked any authority to do anything. But, in the new form of consciousness that had been astroturfed by Lenin and his Vanguard, anything could be true that the victorious proletariat wanted, and, since the Vanguard had proletarian consciousness, and Lenin spoke for the Vanguard, Lenin and his Party had the flexibility to renounce all their former policies without admitting error.

This was way too much flexibility. If the "Revolution" could be democratic, liberatory, pacifist, and socialistic when Lenin's party said it was, and dictatorial, subjugating, bellicose, and market-economic when Lenin's party said it was, it could be anything Lenin's Party said it was. As a result, the "revolution" could freely create new realities (as Karl Rove has put it) and the same verbiage Lenin had used to bypass his Marxist theoretical commitments for the sake of the citizenry, could then be used by Stalin to bypass any commitments to the good of the citizenry for the sake of Stalin's vicious whims.

Under Stalin's tenure, the Soviet Union went from leading opposition to Nazi Germany (the Popular Front) to allying with Nazi Germany (the Molotov-Ribbentrop Pact) to fighting Nazi Germany (the Great Patriotic War) and was dead right each time. (Non-Stalinist leftists used to ask mockingly "What's the Party Line this week, comrade?")

Astroturfing and Psychosis

Psychosis is defined as a failure in ability to distinguish between what's real and what's only in the mind. Insofar as organizations have mental attitudes, and insofar as the Communist Party was committed to the thesis that whatever they wanted was real, they were committed to a psychotic thesis and determined to act accordingly.

I do not want to imply that everyone who took part in any movement they astroturfed is psychotic, or that none of their motivations were real. To the contrary, astroturfers may prefer to manipulate peoples' real pain, and exploit their real exploitation. It makes the fakes more realistic. I do think that the Party collectively had some psychotic ideation, which tended to give way to full-blown psychotic episodes at critical moments. But part of this ideation was that they had started a universal revolution, so their psychotic ideation demanded that it be spread everywhere, to everyone. An example might help.

I knew someone who took part in a radical protest march in the early 1980s. He told me that the protesters were trained in how to behave by the organizers before the march. Much of this was instruction in how to maximize the attention they got while minimizing the punishment, but a lot of time was spent on warning marchers against letting Communists hijack the protest. For example, they were told that they shouldn't allow large gaps to form between groups of protesters, because, once before, Communists had exploited such a gap to put crowd-control barriers in an intersection and tell the following protesters that plans had changed and they should follow them to a different site, where, I guess, they were regaled with panegyrics to Brezhnev.

This illustrates how completely deceit and manipulation had become ends in themselves to the Communists of the late twentieth century. Normal people would never have bothered with such a blatant cheap trick. The cozened protesters would only have listened resentfully and left with a worse opinion of Communism. But to the Communists, the mere act of astroturfing, however lame, counted as a win. The fact that they made non-Communists briefly do something superficially Communistic proved to them they were the astroturfing masters of history they wanted to be. It also illustrates another point: although they were a party of the Left, and the Left tolerated them far too long, by the latter part of the twentieth century, many on the Left had gotten wise to their tricks, and had come to resent them, and learned how to fight them. If the Communists wanted to keep their self-image as leaders of all humanity into the new era, they had to find new people to astroturf. And, as it became harder to do it to the Left, they'd look for easier prey on the Right, unsuspecting and untested in that kind of combat.

What I've just written would have seemed like a paranoid fantasy just three years ago. But the last presidential election has shown beyond a reasonable doubt that there are people nominally on the American Right who are willing to collaborate with the KGB against their domestic political enemies. (Even if, contrary to all the evidence, Russian spies never even tried to help Trump win, the howls of laughter and roars of approval that greeted Trump's invitation to Russia to find and leak American secrets, and the way in which so many "conservatives" supported him even after credible allegations of collusion arose when they could have replaced him with the known quantity, Mike Pence, prove they prefer Russian spymasters to real Americans.) There is no longer any question whether any "conservatives" could betray their country like this, just one of when they started.

With that in mind, I intend to show that the first historically important conservative movement, the Draft Goldwater movement, might have been astroturfed by Russian operatives, starting with and emphasizing how knowledge of past astroturfed "conservatism" can help us understand its current, Trump, phase. Although liberal grassroots movements were astroturfed in that same time-frame (I gave a minor example three paragraphs back, and there's proof that Russian trolls supported anti-Hillary leftists and incited left-wing groups to riot) the focus on the Right is justified since they put a KGB asset in the Oval Office.

From Communist to "Conservative"

Through the 1930s and 1940s, the Communist Party of the United States of America (CPUSA) had a large number of permanent employees on its staff, and the staffs of its district offices. They held various titles, but their real job was activism, and their pay came, not out of member dues, but from Moscow. (That the Soviets supported American Communism, can be shown from Russian Communist documents declassified after the fall of the Soviet Union, the testimony of former American Communists, and the memoirs of FBI double agents Jack and Morris Childs of Operation Solo.) These professional astroturfers gave the CPUSA an important advantage over larger and more deserving radical parties; while the latter had to work to lead the grassroots, the former had trained, experienced professionals to con people that they were their leaders. These professionals were called Cadre or Vanguard, and there were others like them in Communist front organizations, or employed in Soviet organizations such as TASS.

There are no reliable figures for the number of cadre nation-wide, but declassified Party documents say the randomly-chosen Minnesota District supported twenty-six permanent paid staff members with a membership of approximately 800 and Newark supported 18-20 with a membership of 130. Harvey Klehr also cites a private letter that said that Philadelphia supported 16 with a membership of a few hundred (*The Soviet World of American Communism*, p. 163). In comparison, the Pennsylvania Republican Party usually has about twenty permanent full-time staff for 3.2 million registered Republican voters. Extrapolating from these figures, given that the CPUSA membership in 1950 was about 60,000, we could expect three thousand cadre nation-wide or more. I can't find exact major party membership numbers for 1950, but 64,000,000 total is a high educated guess. Extrapolating from Pennsylvania figures, the major parties had a total of about four hundred paid staff. This may be low, but even if we quintuple it, paid Communist activists greatly out-numbered Republican and Democratic ones combined.

I am drawn to the conclusion that the American Communist Party was really just a Communist front—a fake political party created to provide cover for Communist activists.

The pretense that the rank-and-file supported such rela-tively large staffs with their dues had always been implausible, but in the 1950s it became untenable. With the Red Hunt, intelligence breakthroughs, and Krushchev's admission that what people had said about Stalin was true, CPUSA member-ship dropped eighty percent, and about fifteen percent of those who remained were FBI informers. If the cadre had stayed in the Party, they'd have been ratted out, and the FBI would have taken out a lot of Party districts.

No one knows what happened to them instead. Seemingly, thousands of people who lived by drawing attention disap-peared without drawing attention. There's no evidence of mass defections, emigration, arrests, or killings, and the odds of peo-ple with no work experience except Communist activism get-ting other jobs in McCarthy's USA were poor.

Most likely, the Kremlin kept their cadre at work astroturf-ing. But they couldn't openly advocate for Communism, and most others on the Left knew them too well to be pushovers.

About the same time that the CPUSA was imploding, a new, ostensibly grassroots, movement was starting. Though profess-edly and vocally anti-Communist, its leaders (including many professed, vocal, ex-Communists) tended to describe commu-nism as one manifestation of some larger conspiracy or "estab-lishment." This conspiratorial establishment consisted of rich international financiers (in Marxist terms, bourgeois) who were loyal only to their class (as Marxists said bourgeois were) and who monopolized all positions of power and influence in the government and the culture (as Marxists said bourgeois did) and used their power to oppress ordinary working Americans, especially industrial laborers (whom Marxists called proletarians).

The Movement claimed that, since the Establishment were the real enemy, they should be the main target rather than their Communist puppets, who would become harmless once the Establishment were destroyed. Thus, these "anti-Communist" Americans directed their efforts not against foreign communists, but against American capitalists, and, in effect, against American and Western leaders in the fight against Communism. Americans who had experience with Communist attempts to take over American organizations noticed that they used the same tactics, practiced the same sort of discipline, and were organized like their old enemies, and followed ideologies with flaws like those of the Communists they said they opposed. Books (such as *The Radical Right* by Daniel Bell) and articles (such as "The Paranoid Style in American Poliltics" by Hofstadter) stressing these points were writtten at the time, and they were stated on the floor of the Senate (by George McGovern, long before he ran for president). Although there was and is no obvious non-Communist source of skilled activists for this movement, no one serious suggested then that the movement might actually *be* a Communist operation, although many were alarmed by the similarities.

These writers and speakers, mostly liberal or leftist, named them the "Radical Right." They called themselves the "Conservative movement." The John Birch Society was an extreme, and therefor clear, case. While claiming to be more anti-Communist than anyone else, they explicitly held that the *internal* danger to America (America's leaders) was greater than the *external* danger (Communist nations). Once inducted into their secret society, members were told that Communism was just part of a larger conspiracy, sometimes identified with the Illuminati. Strict obedience to Chairman Welch (their unelected leader) was required, and the study of his writings took up most of their secret meetings. They used recognizably Communist tactics to infiltrate and subvert American organizations and institutions (which allegedly had already been infiltrated and subverted by the Conspiracy) and persecuted their enemies even to the point of violent assaults. Krushchev couldn't have asked for better enemies.

For reasons of their own, such "conservatives" tended to claim Barry Goldwater as their leader. The reporter Richard Rovere noticed how anti-American these professed anti-Communists were. Most of the bile regular American conservatives directed against Communism, was turned against American "Easterners" or "Liberals" instead. Communists were given a free pass.

One of the "conservative" groups that sprang up at about this time, supposedly independently, was a faction of the Young Republicans called the Syndicate, led by F. Clifton White, who would become important later on. Clif White and his allies gained control of the Young Republican organization (among others) by using Communistic methods. In White's memoirs, he says he learned how to use Communist tactics from their old enemies, democratic socialists David Dubinsky and Gus Tyler, whom he stayed close to all his life as fellow anti-Communists while fighting a Communist take-over of an organization he belonged to.

White's description of the infighting checks out, except for everything having to do with White himself. No one with personal knowledge of the events mentions White taking any part in any of them including those in which his memoirs have him playing a large role. As for his alleged life-long friends Dubinsky and Tyler, neither mentions him in any published English writings, and Tyler published voluminously in English.

Leninism and the Goldwater Draft

Conservatives often say the grassroots Conservative movement grew out of the Draft Goldwater movement. (Its participants like to call it the "Goldwater Revolution." No intra-party political victory remotely fits the dictionary definition of "revolution," but, then, neither does the paramilitary coup in St. Petersburg in 1917. Such grandiosity is typical of Leninists; everything they do turns out to be "revolutionary.") If so, The "conservative" movement was originally astroturfed by a small group of Young Republican college buddies, led by Clif White.

Clif White was the first Executive Director of what was named the Draft Goldwater Committee after it went public, but was originally called the Suite 3505 Committee by its members (after the address of its Manhattan office.) The members of this committee, who are virtually the only sources for its and White's early activities, gave White most of the credit for Goldwater's nomination for president. He created the rare successful grassroots movement from scratch, starting, they say, with few advance workers (and unreliable ones, according to his complaints.) He also created the only successful political draft movement in US history, despite Goldwater's personal distrust of him, and repeated refusal to run. In the beginning, White had to ask millionaires for political donations without saying whom they were backing (since Goldwater kept refusing to run), or how the money would be spent, but eventually he awakened so much popular enthusiasm that, for the first time, a Republican candidate received more money from small donors than the Democratic candidate. He created his mass movement so quickly and quietly that he routed the party leadership before they could put up a fight. And he did all this from a two-room suite, with only one secretary, using only his own talents and resources, and his consciousness of the grassroots mind.

(White's Committee had little respect for Goldwater as an individual, and he more than reciprocated. They regarded him merely as a vehicle to advance their movement, and he resented their attempts to use him. When he finally got nominated anyway, he largely excluded Committee members from positions in his campaign. White was relegated to a fundraising committee on the periphery. Many prominent "conservatives" have claimed that the Goldwater Draft gave them and many others their start in political activism. If so, it was against Goldwater's will and despite their miserable failure.)

It's been said that the grassroots needed to be enfranchised (have their consciousness raised) by professional activists (the Vanguard) since the "Establishment" kept them too disenfranchised (kept their consciousness too low) for them to act on their own. This is just Leninism with an American face. It's also misleading: the organizers never seriously tried to get grassroots support. They mostly used their resources in states without primaries, where they won most of their delegates by exploiting loopholes in the processes for choosing delegates. The only state where pro-Goldwater forces had a ground game was California. State organizations provided it without help from Clif White's committee. Goldwater barely won against the scandal-tarnished Nelson Rockefeller there, and probably would have lost if there hadn't been a public reminder of the scandal a few days before the election.

This brings us to the next point. In spite of the oft-repeated claim that the Draft Goldwater campaign established a network of grassroots conservative organizations that later came to dominate the political scene, it's hard to find real evidence of grassroots conservatism for years after. By far the biggest factor in Goldwater's nomination was the scandal that destroyed the chances of Nelson Rockefeller, the prohibitive favorite. That left Goldwater as the only candidate of stature, and, as the saying goes, you can't beat somebody with nobody. Even so, the various moderate nobodies that tried to stop him at one time or another got a total of over sixty percent of the primary vote to less than forty percent for him, and the numbers would have been even more lopsided if Goldwater's home state of Arizona hadn't held a primary. Near the time of the Convention, one poll said that sixty percent of Republicans would have preferred Scranton to Goldwater. Polls can be wrong, but it would have taken a ten-point swing for Goldwater to have been even with a Liberal Republican at the time when, according to lore, grassroots conservatives took over the Party. Polls can't be *that* wrong. And then, Goldwater sustained the biggest loss in a presidential election up to that time.

After that, all Goldwater's grassroots support quickly vanished. By 1968, public support for Goldwater was below two percent, while Nelson Rockefeller was still in double digits. In 1968, the "conservative" candidate Reagan got a slightly lower percentage of primary votes than Goldwater had four years before, but it would have been considerably lower had anyone run against him in his very large home state of California. (Besides, Reagan was a better candidate than Goldwater.) The victory finally went to a non-"conservative," Richard Nixon. In 1972, John Ashbrook, one of the three founders of the Draft Goldwater Movement which had made the "Goldwater Revolution," ran as a conservative alternative to Nixon, and never got ten percent in any primary. In New Hampshire, a liberal anti-war Republican ran against Nixon too, got over twice as many votes as Ashbrook, and immediately dropped out because he couldn't win.

Those who argue for grassroots support for Goldwater have to appeal to things besides elections and polling, such as enthusiasm generated, popularity of campaign literature, and small donations (all of which can be astroturfed by trained operatives.)

Goldwater did attract loud, hysterical crowds. If popularity is measured by the followers' excesses, then Goldwater was very popular. Popularity, however, is actually measured by the followers' numbers, and many of the facts advanced to show Goldwater's supporters were numerous, only show they were zealous or well-organized. For example, it's reported that at an early demonstration in New York, three thousand "conservatives" showed up and another three thousand had to be turned away. This sounds like a lot until you remember that New York had about 7,800,000 citizens then, the organizers could have booked a larger venue if they chose, and gathering large loud crowds was a basic cadre skill. At any rate, the crowds disappeared once the votes were counted.

Goldwater's support is also judged to have been large because of the numbers of pro-Goldwater and anti-Johnson books sold. We will spend some time on these, because similar techniques and tactics have been used ever since in "conservative" messaging, even though they had never been used before, nor successfully used by anyone else since, for selling anything. We shall focus on the most famous and best-selling one, Phyllis Schlafly's *A Choice not an Echo*.

Schlafly followed a marketing strategy which didn't involve marketing. She didn't seek a publisher's help, and she didn't place a single ad. By her account, she sent copies to friends with letters asking them to find rich "angels" who'd buy lots of books and give them away.

If you received an unsolicited book from a friend in the mail with a request that you find someone rich to buy large amounts to give away because the Eastern Elites were conspiring to suppress it, would you spend time looking? Do you even know millionaires? But if that's how the book was disseminated, it was astroturfing. What was presented as grassroots popularity was the work of a few wealthy political activists. It is known that more than half of the books were sold in quantities of a hundred or more.

It might be objected that what's important is that when the grassroots got the book they read it and loved it. But there's little reason to believe they read it or loved it. Free literature often goes unread. (How many people read Jehovah's Witnesses pamphlets?) Certainly, some people read it with enthusiasm (or pretended to) but the book wasn't written for a wide readership. The book is an unloveable, Anti-American, Anti-Republican, and vaguely pro-Axis, paranoid screed. It said that, since 1936 at least, American presidential elections had been shams, and the Republican party had been run by a cabal of "Eastern Establishment" "Kingmakers," who, with their counterparts in the Democratic Party, hand-picked winners of rigged elections. (Lenin, in a *Pravda* article called "The Results and Significance of the US Presidential Elections" (1912) described American democracy as a series of "spectacular and meaningless duels between two bourgeois parties," led by "astute multimillionaires" exploiting the American proletariat. Schlafly agreed.)

Schlafly's book is harder on American "elites" than on our

enemies in World War II. In her chapter on the 1940 election she clearly implies that opposing Germany before Pearl Harbor was not only wrong, but unprincipled and disloyal; and Pearl Harbor is never mentioned except to say it was Roosevelt's fault. Communism likewise escapes criticism. A textsearch shows that Schlafly only brings it up to attack American and Republican leadership against it. The foreign "enemy" she spends the most space on is not the Axis nor the Warsaw Pact, but the Bilderberg Group, made up of Western Europeans, our allies. She relentlessly denounces American leaders' and American allies' betrayal of America to Communism; Communism itself doesn't agitate her much.

In 1964 there were still a lot of Americans alive who'd been shot at by Germans and Japanese, and, if many had read the book, it would have been denounced in American Legion and VFW Halls throughout America, and the mainstream media, who had once linked Goldwater to Naziism without good reason, would have been glad to report it. Also, that long before Steve Bannon (who reportedly called himself a Leninist), proud Republicans would have been outraged to be called the dupes of a sham democracy. I suspect most copies went unread, and were printed just to give "conservatism" a spurious reputation for grassroots popularity, and launder illegal campaign contributions as book purchases. (In one letter, Schlafly asks her rich supporters to buy five copies themselves at one dollar apiece, resell them for three, and kick back the two dollars. How many plutocrats would sqander their time pitching the book to five suckers, and then cut the check, when they could save themselves the humiliation and just cut the check?)

Finally, Schlafly stopped printing the book immediately after the election. Printing was only resumed fifty years later, by Rand Paul (it seems Schlafly hadn't kept the copyright up to date). If there'd been real demand for it before Goldwater's defeat, it would still have been profitable after it, and any good capitalist would have continued publishing it. If it had successfully persuaded voters, any good democratic political activist would have continued printing it. Schlafly never even tried; the book vanished the way Goldwater's support did; as if someone had flipped a switch.

Lastly, let's turn to reports that many small donors supported Goldwater. Goldwater operatives bragged about receiving letters from nameless little old ladies containing their rent and grocery money, or full of crumpled dollar bills from anonymous retired steel-workers, who would rather starve and freeze at once than live in a country without President

Goldwater. They seemed proud to have taken poor people's life savings, and possibly lives. But we need not deal on the morality of it, because there is no solid evidence such things happened. The accounts are so vague, they barely count as anecdotes. And, while the total funding Goldwater received from small donors is reported to have exceeded that raised by the Democratic candidate for the first time, it is impossible to verify the reports, since there wasn't (and isn't) any legal obligation to report the identities of small donors. If a million dollars appears in a campaign's coffers and campaign accountants say it's small donations, there's no way to check. What's more, increases in money from Republican small donors occurred *only* in political committees directed by F. Clifton White, of the original Draft Goldwater Committee. If the grassroots had really supported Goldwater, some of them would have sent donations to other Republican Committees, if only by accident. Anyhow, by the next presidential election, four years later, all the small donors White recruited were gone. Republican money raised in small donations were back at pre-Goldwater levels.

What's more, if lots of paranoid, senile poor white people had handed their savings over to Goldwaterites, there would be some mention in the mainstream media of large numbers of demented conservatives sleeping in the streets. Even if the donors were loyal to the death, loved ones would have sued on their behalf, or called the police or the press. And some Attorney General would have made his career by busting Conservative scams.

If "Goldwater conservatives" were willing to let people believe they were stealing little old ladies' last pennies, it must be because they were hiding something worse. And what's worse than that?

One thread running through the historical record is the gross, gratuitous offensiveness of Goldwater supporters, of a kind that is hard to reconcile with confidence in normal democratic processes. It's known that a "Conservative" group in New Jersey called themselves the Ratfinks and passed the time making up songs about machine-gunning Jews in Nazi Germany and suchlike. It is reported that scabrous disrespect was shown to real and imagined enemies by speakers at rallies. Much Goldwater campaign literature was mere libels on respectable American traditions, individuals, and institutions. A Black delegate to the Republican National Convention of 1964 told a reporter a crowd of young Goldwaterites took turns putting their cigarettes out on his new suit. And Rockefeller was subjected to unprecedented abuse before and during his traditional final speech, even as it was being televised.

All of this is the opposite of normal political behavior for Americans or the citizens of any other democracy (let alone human behavior.) Normal democratic politicians understand that electoral victories, like electoral defeats, are never permanent; sooner or later you'll face each other again, and, eventually, you'll be ruled by the people you rule now. Goldwaterites and their heirs act like their victories will be final (like the Great Proletarian Revolution) and the opponents will be destroyed forever (perhaps jailed, or purged). There was much vilification of their opponents, often false, and more often misleading. (Why not kick them on the way up if you're not coming down?) They didn't try to win votes over, but to drive them away from their rivals. (Soviet propagandists were always better at attacking our country than at promoting theirs.) Their victory was won by exploiting loopholes in rules and regulations. (Abusing bylaws and parliamentary rules was standard procedure for CPUSA Cadre.) Their first win was spectacular, but within months they'd badly alienated the electorate. (The Bolshevik revolution excited great enthusiasm. Then reality set in.) And once that happened they dropped the leader they had idolized so recently like he was never really theirs. ("Stalin wasn't truly Leninist.")

In short the extreme behavior of Goldwaterites resembles that of Leninists with no experience in or grasp of democratic politics, more than anything traditional in pre-1960 American politics. Marxist-Leninists held that bourgeois society was full of conflict and suffering (contradictions) even if the combatants had no idea they were fighting and the sufferers never felt pain. The goal of Leninist agitprop was to make these contradictions conscious, so they could be transcended in a Marxist-Leninist utopia. So, the more people fought and felt exploited, the more Leninists won (by Leninist rules). In Lenin's own words:

> The . . . enemy can be vanquished only by . . . the most thorough, careful, attentive, skillful and obligatory use of any, even the smallest, rift between the enemies, any conflict of interests among the bourgeoisie of the various countries and among the various groups or types of bourgeoisie within the various countries, and also by taking advantage of any, even the smallest, opportunity of winning a mass ally, even though this ally is temporary, vacillating, unstable, unreliable and conditional. . . . Those who have not proved . . . their ability to apply this truth in practice have not yet learned to help the revolutionary class in its struggle to emancipate all toiling humanity from the exploiters.

Other explanations for "conservative" combativeness are deficient. "Conservatives" often say they'd been driven to behave despicably by "Establishment" oppression, but it's hard to see how that's even possible. Many of the most combative "conservatives" were too young to have suffered meaningful oppression before the movement started. The New Left barely existed before the Draft Goldwater movement, and can't have provoked them. The mainstream media did treat Goldwaterites badly once they became important, but Goldwaterite vileness was evident before then. The very first issue of the Draft Goldwater Newsletter begins with a tasteless joke about John F. Kennedy being killed (before the assassination,) and then quickly segues into a rant about how Kennedy manufactured crises, including the Cuban Missile Crisis. If anything, screeds like this drove the mainstream media to irresponsible attacks, not vice versa. Memoirs of "conservatives" complain of how they were disagreed with, patronized, disdained, neglected, and occasionally downright insulted. If no one were ever oppressed worse than that, the world would be a better place.

But it was standard Leninist procedure to present minor personal slights as manifestations of intolerable all-pervasive oppression, which could only be defeated by fighting, which was bound to be successful, because it was. And, in a country without major social conflicts that largely agreed on principles and values, they'd have picked fights over those principles and values to astroturf movements that attacked or undermined the consensus (and some "Conservatives" explicitly took the "American consensus" for an enemy.)

Vanguard Conservatism

The 1964 election results showed that "Goldwater conservatism" wasn't really popular. Many of those responsible for the fiasco were driven out of Republican Party leadership positions. President Johnson took their landslide loss as a mandate to enact liberal "Great Society" programs. They were denigrated by the larger culture. (This explains their virulent hatred of American politics, parties, government, and culture.) But they succeeded in one crucial respect: they convinced others they were an American Republican Conservative grassroots movement: they astroturfed themselves. So, by aping them, Vanguard Conservatives could blend in with real Americans, real Republicans, and real conservatives. (This chameleon-like behavior makes it difficult to know exactly who was Vanguard at any time.) They were accepted as

American, but rejected American traditions, because they weren't American.

They overmastered the conservative activist and propagandist classes, because that's what cadre were. Vanguard conservatives, though nominally conservative, actually saw themselves as heroes in an imaginary, roughly Marxist-Hegelian, cosmic historical drama. They were on the right side of History (as Marx said) and could only lose if they thought they had (as Rumsfeld said). Their abject failure was really a success; the world just hadn't realized it yet. They had created a grassroots network of organizations and supporters that was bound to prevail over the long run. And all their dirty tricks were justified, and more would be later, to finish freeing America from the "Establishment," the "Kingmakers," or the "Cultural Elite" (later generations would add the "New World Order," the "Deep State," and the "Reptilians") and make it great again.

They fight for Ideas, but those ideas can become their opposites as easily as vocally supporting the FBI against the KGB became vocally supporting the KGB against the FBI. Like the Russian Vanguard, their rule is to fight ceaselessly until victory, but, because they continually redefine "victory," the fight is always successful and neverending.

Vanguard conservatives have been fighting and beating the mainstream media continually for over fifty years. Their uninterrupted string of victories shows no sign of stopping. Ever.

15
Elders and Brothers

RON HIRSCHBEIN AND AMIN ASFARI

American politics has often been the arena for angry minds. . . . I call it the paranoid style simply because no other word adequately evokes the sense of heated exaggeration, suspiciousness, and conspiratorial fantasy that I have in mind.

—RICHARD HOFSTADTER

M eet the Elders and the Brothers. The Jewish Elders of Zion—a perennial favorite of conspiracy theorists everywhere—thrive in the theorists' paranoid imagination—nowhere else. It's more ominous than you imagine—unless you're one of the cognoscenti: The Elders conspire to strip you of your rights and dignity—if you're a goy (gentile). Just ask the "fine people," the neo-Nazis who terrorized Charlottesville.

These theorists are not fond of Jews *or* Muslims—they apply the same conspiratorial template to both groups. They obsess about a newfound conspiracy that emerged post-9/11—the Muslim Brotherhood. The Brotherhood is real, and it competes with the Elders for the theorists' malign ministrations. Unlike diversity dupes, these Islamophobes know that the Brothers form secret cells metastasizing in high places: Only the gullible and dishonest fail to acknowledge that Obama is a closet Muslim. But Patrick Carlineo, a fifty-five-year-old man from New York, knows about the conspiracy. He's been arrested for calling (Muslim) Rep. Ilhan Omar's office accusing her of working for the Brotherhood and threatening to put a bullet in her head.

The Elders and Brothers' ambitions are anything but modest. In the world according to conspiracy theorists, Jews—inspired, if not guided, by a bogus document, *The Protocols of the Elders of Zion*—conspire to rule the world. As planned, they take over international finance, dominate popular media, and even resort to sponsoring caravans of Central American invaders. Never mind that the caravans (according to President Trump)—include Middle Eastern jihadists (the Jews' sworn enemies).

Speaking of jihadists, we hear a familiar story: The Brothers also seek planetary domination. They conspire to usher in a global caliphate—tyranny that would be over the top—even for the dictator with the bad haircut from North Korea. Such dire prospects spark heated excitement—a fix for adrenalin junkies. It's sooo dull to live in reality. What a letdown if the junkies return from South Neptune to planet Earth: It would be disheartening to realize that the Brothers are about as threatening as their local Elks Club.

As we've already seen, Jews and Muslims have more in common than most people think: both are stigmatized by comparable conspiracy theories—the effluvia of discontent that puts both groups in harm's way in synagogues and mosques. A troubling question arises after we move in for a closer look at each theory: What's their fatal attraction? Strange as it seems, conspiracy theory appeals to the intellect—Gasp! (Before you stop reading, we hasten to add: conspiracy theory is logical *but false*.) Of course, no one lives by intellect alone. Conspiracy theory also appeals to something more powerful and seductive: *Adherents get to live the dream.*

Long Ago in a Prague Cemetery Far Away . . .

When talk turns to worldwide conspiracies, the long shadow of the *Protocols of the Learned Elders of Zion*, purportedly a blueprint for Jewish world domination, can never be far away . . . Donald J. Trump, the Republican presidential candidate, said that his Democratic opponent, Hillary Clinton, "meets in secret with international banks to plot the destruction of U.S. sovereignty in order to enrich these global financial powers" (David Dunlap, *New York Times*, October 27th 2016).

Let's venture into the fevered imagination of the anti-Jewish bigots who spread a Russian forgery about Jewish leaders conspiring in the darkness in a cemetery. Some versions find the Elders conspiring at a Zionist conference in Basel Switzerland. The document seems to be an amalgam of various anti-Jewish tracts and memes. Slightly different stories—same take home message: Jews conspire to rule the world by any means necessary—it's there in writing. Jews intend to dispossess the rightful gentile authorities—be afraid, be terribly afraid!

Curiously, the text begins rather innocently with a recitation of "everybody knows" classical political realism—*Homo homini lupus* (man is a wolf to man). To be sure, renowned advocates of realism get no mention and remain blameless. Even so, seventeenth-century philosopher Thomas Hobbes comes to mind: Given human nature, life is "nasty, brutish, and short" without proper leadership. Here's where the *Protocols* turn ugly. In order to assure such leadership—the reign of Jewish Elders—Jews conspire to profane all that is sacred by destroying the existing order. They will make life nasty, brutish, and short, especially for gentiles. This strategy sets the stage for authoritarian rule by Jews—humanity's divinely anointed overseers.

It is easy to see why the fraudulent work gets bad press— gentiles are degraded. "The goyim are bemused with alcoholic liquors; their youth has grown stupid on classicism and from early immorality, into which it has been inducted by our special agents—by tutors, lackeys, and governesses in the houses of the wealthy, by clerks and others, by our women in the places of dissipation frequented by the goyim." (Hmmm, . . . Jewish Elders plant nice Jewish girls in houses of ill-repute to corrupt the gentiles?) A dissipated, hopeless lot, gentiles deserve to die in "a despotism of such magnificent proportions as to be at any moment and in every place in a position to wipe out any goyim who oppose us by deed or word. . . . Each victim on our side is worth in the sight of God a thousand goyim. (Funny thing: Substitute "Jew" for "goyim" and you have Nazi propaganda with its odious consequences.)

Initially published in 1903, *Protocols* supposedly also reveals Jewish tactics: The ensuing chaos—a return to the miserable Hobbesian state of nature where people do what comes naturally—will make the hysterical mob crave Jewish leadership; Jews at last will assume their rightful place as rulers of the world. In order to usher in this Zionist future, Jews are currently (1903) conspiring behind the scenes to eliminate national loyalty and religion—especially Christianity. They continue to do so—consider the "War on Christmas."

Culture warfare will also demoralize the goyim. The sole speaker in *The Protocols*, the senior Elder, says: "Do not suppose for a moment that these . . . are empty words: Think carefully of the successes we arranged for Darwinism, Marxism, Nietzscheism. To us Jews, at any rate, it should be plain to see what a disintegrating importance these directives have had upon the minds of the goyim." (Seriously, did Jews somehow arrange for the notoriety of these nineteenth-century thinkers?) In any case, no worries: How many have even heard of Nietzsche? Marxism still gets bad press; and many still reject evolution.

The *Protocols* traffics in a popular canard: Follow the money. Jews will exert even more control over international finance—think Rothchilds and Soros. Jews allegedly control the press and manufacture public opinion. And, as they have done from time immemorial, Jews persist in corrupting popular culture with their obscene decadence—the *Protocols* reveals the Jew's pornographic memory. As the Elder reveals:

> Through the Press we have gained the power to influence while remaining ourselves in the shade; thanks to the Press we have got the gold in our hands, notwithstanding that we have had to gather it out of oceans of blood and tears.

Those who translate and edit the hoax cannot resist the temptation to highlight the Elders' proclamation in uppercase letters: "THE WEAPONS IN OUR HANDS ARE LIMITLESS AMBITIONS , BURNING GREEDINESS, MERCILESS VENGEANCE , HATREDS AND MALICE."

The lurid legacy of the *Protocols* is well-documented. Initially Russians invoked the *Protocols* to prove that Jews vowed to destroy Christianity (while Russian Orthodox Christians worshiped a Jewish itinerant preacher as God.) In the aftermath of the Communist Revolution Jews were accused of supporting capitalism and—worse yet—undermining the Stalinist state. Jews in the West were accused of promoting Communism and a crime much worse than undermining Christianity—subverting capitalism. Industrialist Henry Ford had five hundred thousand copies printed. Most notorious: The *Protocols* was one of Hitler's favorite reads; he cited the work in *Mein Kampf*. We'd like to think that the hoax is but a relic of an unenlightened, bigoted past. Not so. It's chilling to read favorable reviews on Amazon. A representative sample:

> This is the scariest thing I've ever read, because they [the Jews] are well on their way to achieving their goals in the US and Europe! Who owns the media RIGHT NOW? Who owns the banks RIGHT NOW? They don't want us to talk about this, because talking is required to TAKE GROUP ACTION AGAINST THEIR PLANS.
>
> No one knows who the original author was, but THIS DOCUMENT IS SPOT ON. Russian revolution. German/Bolshevik revolution. The last 50 years they have been gnawing away at the US and Europe. They have paid groups to mess with reviews and online.

FB, Google, and even Reddit is under their manipulation. "It cries out in pain as it strikes you!" We must defend ourselves before it is too late! Think! Organize! Do!

No wonder Nobel Laureate Elie Wiesel laments: "If ever a piece of writing could produce mass hatred, it is this one. . . . This book is about lies and slander."

Imitation as the Lowest Form of Flattery

Hostility to Islam in the modern Christian West has historically gone hand in hand with, has stemmed from the same source, has been nourished at the same stream as antisemitism.

—EDWARD SAID, "Orientalism Reconsidered"

We turn to the influence of another book—gospel in the conspiracy canon: *Eurabia: The Euro-Arab Axis* by Bat Ye'or (pseudonym of Gisele Littman) published in 2009. It's déjà vu all over again. Predictably, the text traffics in falsehoods. Is it racist? You bet your life! But it doesn't merely play the race card—not that the author plays with a full deck. The author has a grandiose agenda: Conflating race and religion, she conjures up a momentous, make-believe history of the world. We learn that the Muslim brotherhood has long plotted planetary dominion—it began in Europe.

Muslims are bad enough; *Arab* Muslims are worse—culprits spinning round the Axis of Evil. Clueless Islamophobes don't realize that Arabs constitute less than fifteen percent of the Muslim world. Even so, the avid conspiracy believer remains impervious to facts and fixated on the stereotypical Arab. As Said argued, there's a common thread between the hatred of Jews and Muslims—conspiracy thinking. Déjà vu indeed: Like the Jews, Arabs are given strange, if not mystical, powers.

Unlike the *Elders of Zion*, however, the Muslim Brotherhood didn't meet in a cemetery; they hatched their plans for global domination in plain sight! Their front was the Euro-Arab Dialogue (EAD): the driving force behind the Islamization of Europe—and soon America. Conspiracy theorists cite "prima facie evidence" without embarrassment: Haven't falafel and hummus sales skyrocketed in the US? Though the EAD—initially established following the 1973 oil crisis, but subsequently quashed in 1979 after the Camp David Accords—didn't have a clear objective; it's nonetheless useful as a launchpad for the Islamization of Europe.

Paranoid ruminations prompt acting-out. Conspiracy thinking drove the Nazis to commit atrocities against the Jews; after all, they discovered the Zionist plot of the conniving race! Hate crimes against Jews are hardly a thing of the past. Arabs once fared better in this country: magicians in long ago Disney cartoons, Arabs become monsters in Islamophobe conspiracies. Bet Ye'or's paranoia gains traction.

We don't doubt that Islamophobe paranoia prompted Bretton Tarrant to massacre worshippers in a New Zealand mosque. Even those who should know better describe the essential character of "the Muslim," as what Bat Ye'or describes as the *Jihad trait*: an indelible feature of the newly conflated racial-religious categorization ascribed to nearly two billion people. Like some space alien, the 'jihad' trait is a virus that remotely controls its Muslim hosts; it is deeply embedded in their psyche—it's their essence.

There are, however, notable differences in the consequences of conspiracy theories regarding the Elders and the Brothers. Not surprisingly, Jews don't invoke the *Protocols* to indict one another—the forgery isn't a big hit in Israel. (That said, as violence at the Western Wall illustrates, Israeli Jews use other pretexts to malign one another.) Ironically, if not tragically, certain Muslim regimes join the West in persecuting alleged Brothers. In the Middle East, apostasy means trouble—dictators resent fine theological points that indict their corruption. And being overly religious is just as bad—you just can't win if you're accused of being a Brother. In both cases, your life is in danger. The alleged machinations of the Brotherhood became the pretext for the infamous Hama massacre in Syria in the 1980s which claimed nearly twenty thousand lives. The vast majority of those killed were innocent. (Amin takes this personally: His father suffered torture in Syrian prisons because he was falsely accused of being a Brother. His only sin—talking politics with friends!) Moving from East to West, not much changes: Muslims are still accused of belonging to the Brotherhood, this time by conspiracy freaks, right-wing radicals, and white supremacists.

You may see the glaring contradictions presented here, but in the era of Trump, the fringe takes center stage. In a 2016 *Washington Post* article titled "How a series of fringe anti-Muslim conspiracy theories went mainstream—via Donald Trump," Abigail Hauslohner recounted the impetus of these fables. The infamous Reverend Terry Jones warned America of the increasing subversion of American life by Muslims—he called on an end to immigration from Muslim-majority nations, and the deportation of illegal aliens from the US—painfully familiar. But the cleric was late to the party. Islamophobes before him, like Bridgette Gabriel (former reporter for Pat Robertson's television show), Frank Gaffney Jr. (former aid in the Reagan administration), Rep. Peter T. King (R-NY), Rep. Michele Bachmann (R-Minn), and many others saw the jihadist signature, the writing on the wall—a really big wall, it's "Yuge." They knew that the Muslim Brothers were covertly working to undermine American society through a series of conferences, mainly CAIR (Council on American-Islamic Relations)—never mind that these conferences are quite public—stay with us.

Those who peddled fictions have much to gain; they become stars in the conspiracy firmament. They might not always believe what they sell, but they stand their ground with their weapon of mass distraction. Those obsessed with conspiracies revel in being distracted from the actual harms they face from elite misfeasance. You probably know the list of *real* problems that include: environmental disaster, constant war, economic ruin, and corporate greed. The Islamophobia machine worked great, and money pours in from many sources outlined in a report by the Center for American Progress entitled "Fear, Inc." which traced the so-called experts, the funders, and policies (anti-Sharia bills such as the prudent legislation that prevented a Muslim takeover of Oklahoma). Talk about conspiracy, the Islamophobes devised a grand scheme: They spent $40 million derived from seven foundations.

The average American who never met a Muslim bought it; nearly fifty percent of Americans polled by the *Washington Post* held very negative views of Muslims. Ironically what Trump dubs the "Fake News media" may actually reinforce his agenda with negative portrayals of Muslims. Unfortunately, popular media seldom highlight an inconvenient truth for Islamophobes: given the number of Muslim healthcare professionals in the US, you're more likely to be healed by a Muslim than killed by one. No "film at eleven": If it doesn't bleed, it doesn't lead!

Terrorism becomes entertainment in the popular media, featuring Arabs as the perpetrators of horror. They're everywhere at once, anywhere in the world that violence occurs. The Arab bogeyman is the sum of all childhood fears—febrile dreams that came in darkness. Turning to entertainment charitably called news, talking head "experts" are rarely Muslim. This isn't the first time such careless—perhaps malicious—coverage of a group took place. Consider the war on drugs: remember the stories about African-American addiction to crack cocaine? Surely the white-privileged renounced all drugs.

Conspiracies work well because they play to our deepest fears, especially our childhood anxiety and insecurity. Such theories allow us to project our insecurities on other groups— our failings are never caused by us; we can't be that inept— must be the "other." Eric Hoffer's *The True Believer: Thoughts on the Nature of Mass Movements* was prophetic—he compels us to consider our love of conspiracy as a way to explain our shortcomings. Whereas developing societies may adhere to paranormal explanations for their failures (magic, jinn, demons), we can't do that—we're not a monolithic society, and we're supposed to be "civilized." So Jews and Muslims are given supernatural powers, even by seemingly rational experts. Such ruminations are not unknown in high places. A leader like Trump enables the conspiracy to thrive; he bestows legitimacy. Thank God he's making us great again!

Meanwhile, back at the Brotherhood, even progressives such as Bill Maher warn that the Muslims have nefarious objectives: assimilating into American culture; slowly gaining power within American institutions; eventually subverting our way of life under the yoke of Sharia law. Don't trust them, even those feigning modernity. There are no "moderates": all are driven by a single, immutable ideology—corrupting Western civilization. Perhaps some of us (Muslims) missed these classes in school—confession, my grades weren't always a source of pride. I don't recall much talk in school among Muslims of plans to ruin the West. Perhaps our greatest sin was trying to convince Americans that falafel is healthy, even though it's deep-fried.

No doubt, those who traffic in conspiracies enjoy some success. By portraying Muslims—especially Arabs—as subversive, no matter their level of "Muslim-ness," conspiracy freaks can justify claims that *anything* and *everything* done by a Muslim is ill intentioned. For example, Elham Manea (a Swiss Yemeni), a member of the European Foundation for Democracy (EFD), a right-wing group that dispenses *advice* to politicians concerning Islamism, stated that Islam deserves relentless suspicion. Even something as benign as a request for prayer rooms by Muslim college students should be seen as nothing less than an Islamist agenda. In Sweden—the supposed bastion of tolerance—far-right groups demanded that Mehmet Kaplan, a former housing minister, should step down due to his supposed ties to "to the Brotherhood."

The pattern is clear: Such conspiracies aim to undermine *any* influence by Jews or Muslims. In both anti-Jewish and anti-Muslim conspiracies, the message is that *these people simply cannot be trusted.* Adherents to these conspiracies feel superior because they *know* the *truth*. A Jew or Muslim who runs for the school PTA or Congress has the same objective: undermining America's culture, bringing the country to ruin, taking away 'American' jobs, or worse yet, their lives. (The *Protocols* redux.) Americans will be subjected to the whims and desires of fanatical Muslims and Jews. Our hapless countrymen will be forced to wear burkas in Chicago, and Kippah's in California, and circumcision will become compulsory at local malls.

Surely the conspiracy is real, as the appearance of new Muslim congresswomen and legislators attests—creating fertile soil for confirmation bias. Speaking of psychology, those who study intergroup relations understand that in multicultural societies, powerful, dominant groups who feel threatened resort to tactics of exclusion. Indeed, Jack Levin's 1975 book *The Functions of Prejudice* reminds us that prejudice is a useful tool for the powerful; it is used to distinguish between members of the in-group and those from the out-group. However, conspiracy theorists are not overly attentive to the psychological literature. The similarities between the two conspiracies are disconcertingly uncanny:

> There exists a subterranean world where pathological fantasies disguised as ideas are churned out by crooks and half-educated fanatics for the benefit of the ignorant and superstitious. There are times when this underworld emerges from the depths and suddenly fascinates, captures, and dominates multitudes of usually sane and responsible people, who thereupon take leave of sanity and responsibility. (Norman Cohn, *Warrant for Genocide*, p. 314)

Norman Cohn got it right, but what makes these conspiracy theorists tick; what's the fatal attraction?

It's Only Logical

Garbage In, Garbage Out

— an early computational meme

Other Jewish bogeymen may haunt the fever dreams of the vicious, but the scale and intensity of the attacks on Soros are unrivalled. They reveal what the global nationalist right believes is at stake in this present moment. We may one day look back on this era as the Soros Age of anti-Semitism.

Conspiracy theories are logical—*but false!* How can this be? *If you accept the premise,* the conclusion follows—*a big "if."*

Consider a crucial distinction, a distinction especially precious to logicians: Valid arguments don't necessarily have true conclusions. Here's an overused example inflicted upon our students.

All dogs are cats
Spot is a dog

Therefore, Spot is a cat

Students who aren't on cellphones or Facebook notice the straightforwardly absurd premise that begins this argument. The take-home message: Arguments can be, structurally, solidly logical *and* have premises or conclusions that are blatantly false. Even so, once upon a time, rationalist philosophers believed that logically coherent arguments in and of themselves revealed the truth about the world—conspiracy theorists and other paranoids still do.

Validity and truth are distinct species. The false premises of conspiracy theory may reflect the received wisdom of a revered authority. But more often than not, these theories are scaffolded on bigotry, wishful thinking, or—if Freud got it right—unconscious, long repressed passions. This syllogism puts a rational face on bigotry; it's intuitively obvious to those living in the deepening shadows of the *Elders of Zion*:

Jews hatch nefarious plots to rule the world.
George Soros is a Jew.

Therefore, George Soros hatches nefarious plots to rule the world.

Muslims, to be sure, are subject to the same impeccable reasoning. Conspiracy theorists cling to such premises as if their life depends upon it—they believe it does! Nothing too complicated about Islamophobe logic:

All Muslims want to destroy American civilization and rule the world
Rep. Ilhan Omar is a Muslim.

[Draw your own conclusion.]

The theorists seek confirmation bias—instances that prove their point. They scrutinize Jews and Muslims, not Unitarians and Presbyterians. And they're oblivious to non-Jewish elite decision-makers in government and industry who may be responsible for their discontent. Long story short: counterexamples are overlooked (cognitive dissonance) or dismissed with contempt—fake news (cognitive insolence). Conspiratorial premises emerge from troubled imaginations, not unbiased observation. There is no conceivable disproof. True by stipulation, the premises cannot be disproved.

Living the Dream

I can't sit by and watch my people get slaughtered. Screw your optics. I'm going in.

—ROBERT BOWER (accused of October 29, 2018 massacre at Pittsburgh synagogue)

Like other tried and untrue conspiracy theories, apocalyptic tales of the Elders and Brothers are mythic. Conspiracy theory is a dream-like myth, a myth Joseph Campbell called a hero's journey. The would-be hero's familiar world implodes; he endures the dark night of the soul. And yet, there's transcendence and promised redemption. As if by magic, the cowering victim of modernity and changing demographics shape-shifts into a soldier destined to defeat his people's mortal enemy by any means necessary—preferably heart-stopping violence. Triumphing against formidable odds he wins the tribe's adulation and attains ersatz immortality.

This heroic narrative denatures a confusing, troubling world of ambiguity. Theorists embrace the "KISS Principle" (keep it simple, stupid). Jews and Muslims are reduced to seamless wholes—alien tribes of one mind. Men like Robert Bower and Bretton Tarrant (the alleged perpetrator of the New Zealand massacre) see themselves as soldiers in a cartoon world in which luminous forces of light are locked in mortal combat with the darkest principalities of evil. In a supreme act of courage, Bower virtue-signaled and boasted he was "going in" to battle demons—innocent, unarmed worshippers.

Bower and Tarrant fantasized about a glorious future, but reveled in the here and now as they discharged their weapons. This discharge (Freud read sex and violence into our every move) evoked what the Father of Psychoanalysis deemed the greatest pleasure: the sudden release of long-repressed instinctive passion. Conspiracy theorists ordinarily get by through repressing homicidal rage, but like Mick Jagger they can't get no satisfaction. But these are extraordinary times. Enacting the myth, the wannabe hero in Campbell's words, overcomes the monotony and humiliation of everyday life and "and actually feels the rapture of being alive." To paraphrase philosopher William James, it does no good to indict the horror of mass murder; the horror *is* the attraction.

Could it be that the conspiracy theorists don't realize that they've pushed too far? An irony emerges as Jews and Muslims push back: Prejudice also serves the excluded group's interests; it encourages camaraderie with other persecuted groups. We're heartened to see enmity turn to newfound amity as Jews and Muslims join to push back. We Jews and Muslims realize we're all in this together.

PART IV

"The trouble with conspiracies is that they rot internally."

16
The Defining Conspiracy

JAMES ROCHA AND MONA ROCHA

SOCRATES: Oh, if it is not my good friend, Conspiratos!

CONSPIRATOS: Why, Socrates, my dear friend! What brings you to the agora?

SOCRATES: Why, I am not here to corrupt the youth, if that is what you are implying! [*Both laugh heartily.*] In fact, I am just here to visit with friends and acquire wisdom. And how fortune shines on me to find a person with whom I can accomplish both tasks! What is it you are up to, my gentle Conspiratos? Why are you perchance at this fine wine establishment? Are you writing poetry? Perhaps a play? Maybe you are here working on a blockbuster screenplay?

CONSPIRATOS: None of these vain pursuits, Socrates. I am in fact working on something much more noble: I am arguing with someone on the Internet!

SOCRATES: Ah! I have heard amazing tales of this "Internet" where only the highest level of intellectual debate occurs. Surely you must tell me of the great wisdom you are acquiring from your interlocutor!

CONSPIRATOS: While many of the tales you have heard of the Internet are indeed true—it is a remarkable place where one can just as easily share cat memes as discover sage philosophical treatises, largely on the unconditional worth of the cat memes—that, unfortunately, is not my current quest. Instead, Socrates, I am arguing with a conspiracy theorist!

SOCRATES: By Hades! Say it is not so, Conspiratos! Is someone actually wrongly attributing an overly simplified causal theory for a significant event? And doing so on the Internet of all places? I have yet to encounter this use of the Internet myself, though I have indeed been quite impressed by the aforementioned cat memes.

CONSPIRATOS: I am afraid it is so, Socrates. There are not only wonderful people sharing the cutest pictures of their cats, which are then enhanced by the wittiest of scribes writing as if they are these very cats speaking in a human tongue, but there are also people on the Internet who propose wrongful theories of causation. That is, Socrates, there are conspiracy theorists on the Internet. Unfortunately for them (and fortunately for the rest of the world), I have arrived just in the nick of time to prove them wrong!

SOCRATES: Well, this surely is a shocking turn of events— surely this is not what the Internet was originally intended for and none of its founding creators, such as Algoreos, could have ever imagined this unintended consequence! Yet, my good friend, Conspiratos, I feel we may have, perchance, happened upon a felicitous circumstance!

CONSPIRATOS: That would be good news indeed! How, please tell me, have the Fates rewarded my hard work here?

SOCRATES: Well, I'm glad you asked, Conspiratos, because I am certain that the Fates are actually rewarding me, but I know not why. For you see, I am about to learn what a conspiracy is such that I can truly comprehend why all conspiracy theories are false. I am the one who has the good fortune of happening upon someone who surely holds onto this wisdom and can now share it with me. Thank Athena herself for this wondrous occasion!

CONSPIRATOS: [*after a hearty laugh*] Why, Socrates, you ask for so little! Of course I would be happy to answer your questions.

Just then, a barista calls out, "Socrates," and he goes to claim his glass of wine, rightly mixed with water. Socrates returns to the table, pulls up a chair, and prepares to learn what a conspiracy is.

CONSPIRATOS: Well, my dear friend, nothing could be simpler than to explain to you what a conspiracy is, and to even ensure that the explanation shines light on why conspiracy theories are always wrong. A conspiracy is just like what I am arguing against now: the idea that a group of reptilian gods who control the government have installed Trumpes as our leader so that he can increase the value of Twitter stock by leading so many people to take to Twitter just to angrily respond to his tweets.

SOCRATES: By Zeus, a conspiracy sure is a complicated thing! I truly appreciate your hard work in arguing with anyone on the Internet who is supporting such a wild conspiracy.

CONSPIRATOS: I do it out of duty, Socrates. It simply must be done.

SOCRATES: I see; I see. But here is my issue, Conspiratos, I too want to argue with simple-minded people on the Internet.

CONSPIRATOS: And you would be very good at it!

SOCRATES: Why, thank you. I appreciate your kind words. But my problem, Conspiratos, is that you have merely given me one example of a conspiracy. I would love to go argue with others spreading conspiracies, but I have yet to receive a definition that I can apply to other cases.

CONSPIRATOS: Ah, I see, Socrates. Yours is a fair request indeed. I shall go ahead and provide you with a definition that will enable you to argue with idiots on the Internet in a wide variety of situations.

SOCRATES: I would be very gracious for such assistance, indeed, Conspiratos!

CONSPIRATOS: Well, legally speaking, a conspiracy is two or more people who work together in secret to plot some evil deed.

SOCRATES: Well, I am, as predicted, grateful that you have provided a definition that can apply to a plethora of situations. Yet, I find myself confused, Conspiratos. Won't this definition apply to too many situations?

CONSPIRATOS: Please tell me the basis for your concern, Socrates. Apply to too many situations? How so?

SOCRATES: Well, when we began our quest, you made clear that a conspiracy theory is always wrong, and, as it seemed to me, typically idiotic.

CONSPIRATOS: Oh, yes, Socrates, both statements are entirely true, without exception.

SOCRATES: But Conspiratos, are there not conspiracies as you have just described every day of the week and twice on Saturday?

CONSPIRATOS: By Achilles's heel, Socrates: You are right! A bank robbery, a train robbery, or even a robbery of the temple treasury would all be conspiracies by this definition.

SOCRATES: And there is nothing idiotic about theorizing that the temple was robbed by a group of people who had planned out their evil in advance.

CNSPIRATOS: Most certainly not! I have provided my definition in haste. Let me add the key detail that would make it true.

SOCRATES: Please do.

CONSPIRATOS: A conspiracy is when two or more powerful, rich, or famous people work together in secret to plot some evil deed. You see, Socrates, the fools on the Internet are always yammering on and on about how they are being swindled by the rich and powerful, when in fact it is more often than not simply that conspiracy theorists are plain fools whose drudge in life is their own fault.

SOCRATES: But, my dear friend, are there not many times when alleged conspiracy theories, based on this definition, turn out to be entirely true?

CONSPIRATOS: I think not, Socrates, but please enlighten me if it is so.

SOCRATES: Do you remember the tale of Elizabeth Holmes?

CONSPIRATOS: Why, Socrates, we live in ancient times, but we do not live in the Dark Ages! Of course I have seen the documentary on HBO that explained her various contrivances that duped both investors and patients by pretending that she had invented a nearly magical machine that could analyze over 200 diseases with merely a drop of blood.

SOCRATES: Yes, we have seen the same documentary. Surely Holmes and her partner were rich and powerful people who conspired to deceive people at the risk of those people's very health.

CONSPIRATOS: I see what you, mean, Socrates! Fortunately, I have remembered a significant element that I have failed to mention. When I argue with nimrods on the Internet, they are in fact quite assured that it is the government that conspires against them. Let us say that a conspiracy is when two or more powerful governmental agents work together in secret to plot some evil deed. Surely this definition shall suffice for our purposes!

SOCRATES: Well, I hope you are right. Shall we put it to the test?

CONSPIRATOS: I believe it shall succeed like a great Olympian!

SOCRATES: Let us find out. Can we not imagine scenarios where people theorize that the government is involved in some secret evil plot?

CONSPIRATOS: Why yes, Socrates: that is what happens all the time on the various subreddit threads that are populated by conspiracy theorists! It is the very reason for which we must labor to end this dangerous nonsense peddling!

SOCRATES: Very well. But do not some of these secret government plots eventually come to light?

CONSPIRATOS: I think not, Socrates!

SOCRATES: Well, aren't the Tuskegee experiments exactly the sort of conspiracy you have just here described? The government plotted in secret for over forty years to run experiments on African-American men. Surely, we can refer to this as a secret, governmental plot to engage in evil activity.

CONSPIRATOS: I see what you mean, Socrates.

SOCRATES: And does the government not regularly engage in illegal and evil acts in times of war that they also keep secret from us? Acts that we only discover later?

CONSPIRATOS: Well, surely, Socrates, that happens all the time, from the battle at Troy, where I must admit that Achilles's acts were not always entirely above board, all the way to the wars in Iraq! No honest person can deny that there have been numerous governmental war crimes. I see what you mean, Socrates, and I am grateful for you pointing out my mistake. I believe the issue here is that these are not conspiracies because the government intends well.

SOCRATES: Even in cases of illegal warfare or deceitfully experimenting with the lives of African Americans?

CONSPIRATOS: I believe we both know that we have different political ideologies, Socrates. And this is a matter we surely will not settle here.

SOCRATES: Yes, Conspiratos, I agree. We should seek fruitful agreement instead, if possible.

CONSPIRATOS: Yes, well, can we agree that the government believes itself to be acting justly when it pursues an advanced understanding of syphilis or when it deceives the Trojans?

SOCRATES: Well, Conspiratos, I do not pretend to know the mind of others, and government actions in these cases do not inspire me to be charitable. But I can at least acknowledge that your assessment may be correct for all we know.

CONSPIRATOS: I do believe it is sufficient for our purposes that these are not the kinds of cases I had in mind. I am not interested in cases where people disagree over the government's intentions.

SOCRATES: Then please tell me the kinds of cases that count as conspiracies as you understand the term.

CONSPIRATOS: A conspiracy is when two or more powerful governmental agents work together in secret to plot some significantly evil deed with *clearly* malicious intent. Thus, if some people think the war is justified, they may see war crimes as necessary evils, and not as acts done with malicious intentions. Such cases should not count.

SOCRATES: So, we are looking for cases where pretty much everyone would agree that the evil deeds were significant and done with malicious intention?

CONSPIRATOS: Only those cases.

SOCRATES: So, I wonder, Conspiratos, whether there are cases where we must admit that there are in fact conspiracies of the kind you have just pointed to.

CONSPIRATOS: Socrates, there are most definitely not . . . If you think otherwise, go ahead: Change my mind!

SOCRATES: We know of numerous tyrannical governmental leaders who have worked with their supporters to secretly and maliciously plot evil deeds: Pol Pot, Nicolae Ceausescu, Augusto Pinochet . . . I can go on and on.

CONSPIRATOS: I believe everyone knows you can go on and on by now, Socrates . . .

SOCRATES: I could even name some tyrants from our own time period . . .

CONSPIRATOS: . . . Perhaps it is best if you did not. We do not want you to be put on trial for saying the wrong thing with these various conspiracies going around today . . .

SOCRATES: Ah ha! "Conspiracies"!

CONSPIRATOS: You indeed got me there, Socrates! I just now used the word as if it were a perfectly reasonable word. Clearly, we ought to be more careful.

SOCRATES: And such care will come with a superior definition.

CONSPIRATOS: Agreed. I will seek to repair my previous definition with the addition of merely two words. A conspiracy is when two or more powerful, but democratic government agents work together in secret to plot some significantly evil deed with clearly malicious intent.

SOCRATES: I see what you did there. Is this addition not rather ad hoc?

CONSPIRATOS: What do you mean by that?

SOCRATES: Well, Conspiratos, you are telling me that there are no powerful governmental agents engaging in conspiracies.

CONSPIRATOS: None whatsoever.

SOCRATES: But now you admit that there are many—they just do not happen to rule in democracies.

CONSPIRATOS: Yes, indeed. I see what your objection is, but I would push it to the side. Socrates, what concerns us here is whether people engage in conspiracies. I argue that they never do. Any reasonable person would agree. Now, you have thus far been very helpful in attending to the need to define "conspiracy" so that we can be sure that no one engages in conspiracies. But obviously those of us sane people who think that no one would ever engage in a conspiracy realize that tyrants do so all the time. So clearly, we are not claiming that tyrants would not engage in conspiracies. We are claiming that powerful governmental officials who work within democracies would never engage in conspiracies. It is absurd to think they would. Democratic leaders may sometimes do bad things, but only because they believe them to be in the people's best interests. That is why they sometimes engage in illegal wars (but only to defeat tyrants elsewhere) or experiment on their people (but only to find cures for diseases). But democratic leaders would never maliciously engage in secret plotting to commit significantly evil acts.

SOCRATES: I see what you mean. Thank you for this wonderful clarification. I feel like I have a much better picture of what you are saying is impossible now.

CONSPIRATOS: You are welcome, Socrates. I am glad we came to a point of joint knowledge over the matter.

SOCRATES: Hold on, Conspiratos. I am afraid I know nothing on this matter, just as I stated!

CONSPIRATOS: By Medusa's hair, Socrates, what problem do you have now?

SOCRATES: Is there not widespread oppression in democracies?

CONSPIRATOS: Of course, there is. No one would deny it.

SOCRATES: And do we not consistently find powerful democratic leaders acting in secret to advance this oppression of underprivileged groups?

CONSPIRATOS: While I believe much of oppression is structural in nature, I suppose there are sometimes governmental leaders involved.

SOCRATES: Well, do these democratic leaders from time to time pass bills that they should know would be immoral and even evil since the resulting laws will undeniably enhance oppression?

CONSPIRATOS: Can you give me one example?

SOCRATES: If I could only give just one! What about slavery, Jim Crow, mass incarceration? Look how long it took for women to receive the vote, for marital rape to become illegal, for sexual harassment to be legally actionable, for gay marriage to be legally acceptable and protected? We still wait for discrimination on the basis of sexuality to be actionable at the federal level. Trans people continue to struggle for their rights. There are so many criminal laws whose enforcement show clear patterns of racial discrimination . . .

CONSPIRATOS: Socrates, these cases are surely not secrets.

SOCRATES: If only the relevant government leaders felt the need to keep them secret! Your point was that conspiracies are impossible because powerful democratic leaders would not maliciously do evil deeds. But in furthering oppression, these very people of whom you speak have routinely done evil deeds. More significantly, they often did them in the light of day. I'm sure many more of their oppressive actions were hidden behind closed doors, but the point here is that they were not frightened away from doing evil deeds even when the world was watching. So why should we accept that it is impossible for democratic leaders to engage in acts of evil in secret when they are willing to do so in public?

CONSPIRATOS: Why Socrates, you are sounding rather cynical—I fear your skepticism of the government is starting to make you sound like one of these conspiracy theorists on the Internet . . .

SOCRATES: Please, my dear friend, all I seek is knowledge.

CONSPIRATOS: Well, let me enlighten you. Everyone has personal biases. People are racist, sexist, heterosexist, transphobic, and so forth. No one denies this point. And so, yes, you have caught me: I must admit that even democratic leaders are flawed humans who make mistakes. Look, Socrates, even the Gods make mistakes! Do you think Zeus has no biases when he is deciding whom to hit with a lightning bolt? I cannot expect Congress members to be less biased than Zeus himself!

SOCRATES: But these "mistakes" involve the oppression, sometimes over decades or even centuries, of millions of people! This so-called character flaw is not merely a bias that involves tipping some people better than others (though that too is potentially worrisome). We are talking about severe harm and downright evil!

CONSPIRATOS: Fine, Socrates. Whatever. For the sake of moving on with my life, I will simply grant what you are saying and adjust my definition.

SOCRATES: Please do so.

CONSPIRATOS: A conspiracy is when two or more powerful but democratic leaders work together in secret to plot in a rational fashion (and are not merely giving in to their bigoted biases) some significantly evil deed with clearly malicious intent.

SOCRATES: It sure feels like your definitions are becoming much longer and more ad hoc.

CONSPIRATOS: You're welcome to your opinions, Socrates, even if they sound like those of a child who has lost his dice.

SOCRATES: Come now. Let us try to remain friends and continue to seek knowledge together.

CONSPIRATOS: Your mama seeks knowledge.

SOCRATES: Indeed, she does. We are a very progressive family who believe that women should be educated, much like Spartan women. So, let's return to this definition. Do you not know that Nixon and his aides worked in secret to plot in criminal and malicious ways to destroy their enemies and manipulate elections?

CONSPIRATOS: I do not think anyone doubts that, Socrates.

SOCRATES: And were they not plotting rationally? Even if they had bigoted biases (I do not mean to deny that they did), it was not those biases that motivated them, but plain old self-interest and greed. Was it not?

CONSPIRATOS: I once again see where you are going with this line of thought, Socrates. And, frankly, we could end up going in circles all night with this kind of nonsense.

SOCRATES: Please then set me straight and give me the correct definition.

CONSPIRATOS: I will do just that so I can get back to something truly important: arguing with chumps on the Internet.

SOCRATES: Very well! Please do not hide the final resolution any longer!

CONSPIRATOS: A conspiracy is when two or more powerful, but democratic government agents work together in secret to plot in a rational fashion (and are not merely giving into their bigoted biases) some significantly evil deed with clearly malicious intent, and the government never officially acknowledges that it is true. We are done here, Socrates.

SOCRATES: How so?

CONSPIRATOS: I see what you have been doing all along. You keep using examples of alleged conspiracies that the government has eventually acknowledged were true. Yes, slavery was real. Yes, governments lie in times of war and commit war crimes. Yes, women and the LGBTQ community have been treated horribly. Yes, Nixon was a horrible president. Yes, to all of your charges. I'm surprised you had the decency not to bring up the government spying on political dissidents, such as the Black Panthers and Martin Luther King, Jr., in COINTELPRO—yet another alleged conspiracy that the government now acknowledged happened. Yes, the government did some very bad things, but they all follow a similar pattern: a democratic government did something that meets our prior definitions of conspiracy, it was caught, it was stopped, and then the government admitted it was wrong. These then are not conspiracies. The only conspiracies are the ones that the government won't acknowledge happened because they are patently absurd. And those kind—the absurd conspiracies—never happened, and we know that because the government would come clean once they realized their mistakes.

SOCRATES: I see, Conspiratos. So, a conspiracy is only a conspiracy when the government continues to deny it happened.

CONSPIRATOS: Yes, indeed.

SOCRATES: And before the government acknowledged them, these very cases of past evil deeds, which we now acknowledge are real, were impossible *at the time when the government denied them.*

CONSPIRATOS: Yes, exactly.

SOCRATES: So, your whole case is that there is no such thing as a conspiracy because something is only a conspiracy if the government denies it, and if the government denies it, then we have no way to know it is real, and, thus, it isn't real just because we can't know for sure if it is real?

CONSPIRATOS: You have finally said something that makes sense.

SOCRATES: Conspiratos, it looks as if you have defined conspiracy in such a way that you will simply beg the question and always assume you are right against anyone who disagrees with you. Thus, the conspiracy theorist is always wrong because they cannot give evidence the government accepts, and if they have evidence the government accepts, you will simply deny it is any longer a conspiracy. When it is a conspiracy, it is impossible and doesn't exist. When it is confirmed to be real, then we simply say it isn't a conspiracy.

CONSPIRATOS: Socrates, you are making that sound like a bad thing, but it proves all conspiracy theorists are wrong all of the time.

SOCRATES: I believe that we have only proven here that the only way to define "conspiracy" so that all conspiracy theories are false is to beg the question so that no conspiracy could ever be true. If we attempt to define conspiracy in a way that is not ad hoc or question-begging, we will see that some conspiracies are true while others are false, including some that are even laughably false. Surely the true conspiracies are the less fantastic ones: widespread oppression, war crimes, greedy politicians, and even occasional sex scandals gone wild. It probably is still true that almost all conspiracy theories, especially the large-scale, fantastic ones pushed by people with whom you wish to argue on the Internet, are absolutely false. But it is not that conspiracy theories are all false by definition, unless we cheat in our definition.

CONSPIRATOS: Socrates, this is why you have no friends.

SOCRATES: I have always believed that my lack of friends is the grandest conspiracy of them all!

CONSPIRATOS: 'Bye, hopefully forever. I now wish for the government to conspire to make you drink hemlock.

SOCRATES: 'Bye to you as well, my dear friend! [*Our interlocutors go their separate ways, with Socrates tipping back his wine and taking a hearty swig on his way out, and Conspiratos muttering to himself about Socrates's ability to corrupt everything.*]

17

Should You Be Wearing a Tinfoil Hat?

MICHAEL GOLDSBY AND W. JOHN KOOLAGE

In fiction, conspiracies make the best antagonists. Hydra in Marvel Comics and the associated Marvel Cinematic Universe is behind many of the nefarious plots that the heroes must face. Hydra's reach is broad, even corrupting Steve Rogers (a.k.a. Captain America) in the *Secret Empire* story arc, and their organization is nearly inextinguishable despite the heroes' best efforts.

Other examples include Opus Dei in *The Da Vinci Code*, SMERSH and SPECTRE in the Bond novels and movies, and nearly every faction with at least two people in *Game of Thrones*.

One reason conspiracies make good antagonists is that a conspiracy presents a foe that taken together presents a real challenge for the hero or heroine, even if the protagonist can easily deal with each conspirator individually. Conspiracies make cartoonish villainy somehow more believable.

In fiction, conspiracies tend to make the villainy more believable. On the other hand, hypothesizing a conspiracy in real life seems less credible. It doesn't sound plausible that a respected general engaged in a conspiracy to use cardboard tanks, planes, and jeeps to deceive a supervillain, who is famously prone to fits of apoplectic rage, into thinking that an invasion would come from one direction rather than another. It stretches credibility to think that it happened. Yet in the real world during a real war (World War II) that's exactly what happened.

In real life, when people talk about conspiracies, it's easy to dismiss their theories as a paranoid fantasy. The phrase "take off your tinfoil hat," which finds its roots in a conspiracy theory has become synonymous with "stop being paranoid, your claims are not reasonable." Most people think that any conspiracy theory is a fantasy, not worthy of serious consideration. Still, some conspiracies are true, and some, while entertaining, stretch the most permissive standards of plausibility. What we need is a way to distinguish between these two types.

Conspiracy Theories Tend to Look the Same

Well-designed conspiracy theories tend to look alike. A better way to put it is that they tend to have the same form. Philosophers like to talk about formal similarities, because it allows us to talk broadly and abstractly about many things. With well-designed conspiracy theories, that form includes four key features.

- an agenda or motive,
- a group of conspirators who are pursuing that agenda,
- a plot to achieve that agenda, and
- a cover story that conceals the actions of the conspirators from the general public.

Every conspiracy theory is centered around an *agenda*. Not every agenda-centered activity rises to the rank of a conspiracy, but no conspiracy lacks an agenda. It's the agenda or motive that does much of the "explanatory" heavy lifting for conspiracy theories. The agenda is also heavily cited in conspiracy theories as a way of accounting for why the conspirators bother to take such extreme measures. It often turns out that the agenda forms an important part of the "evidence" cited in support of the conspiracy theory.

Agendas can come in many forms. Some agendas are aimed at profit. A whole class of conspiracy theories assert that a low-cost, clean energy alternative has been developed that the top energy companies have chosen to suppress to maintain high profits. Other agendas are aimed at maintaining power. Conspiracy theories focused on the Kennedy assassinations often suggest that the Kennedys were going to seriously limit the power of the military-industrial complex, which is why the conspirators conspired to kill both JFK and RFK. Some agendas, according to these theories, are aimed at expanding power, which is the agenda attributed to such grandiose conspiracy theories as those involving the Bavarian Illuminati, the

Freemasons, and Skull and Bones. Other motives are possible, but those are the top three.

Every well-designed conspiracy theory involves some *conspiratorial group*. A conspiracy of one is technically possible, on our understanding, but it is unlikely. While Fred may have an agenda of global domination, Fred, by himself, is not really much of a threat. Conspiracy theories often involve many moving pieces. First there is the plot to manage, and a conspiracy has to make sure that the plot is actually making progress toward achieving the agenda. Secondly, the plot must be hidden, so as to avoid resistance and prosecution. This takes considerable resources. The conspiratorial group is what provides those resources. If we think of the agenda as the motive for the conspiracy, the conspiratorial group serves as the means.

The agenda provides the why of the conspiracy, and the conspiratorial group provides the who and what resources they have. *The plot* tells us exactly how the conspirators achieve their agenda. What actions do the conspirators take to achieve their goals? Do they subtly recruit or bribe people in places of influence? Do they bury new technology? Do they perpetrate an elaborate PR campaign to influence public opinion and the powers that be? In the JFK assassination conspiracies, the conspirators recruit or influence a patsy to kill the president before he can take action to limit the power and influence of the military-industrial complex.

In most cases, the plot is either illegal, unethical, or both. At the very least, the optics surrounding the plot would lead to great embarrassment for the conspirators. Additionally, if the plot is discovered, then its chance of success declines. The *conspirators' cover story*, then, is necessary to prevent the discovery of the plot in order to avoid those hazards. The conspirators manipulate evidence or use their connections with the mainstream media to provide the cover story to the public.

The cover story does not portray what is really happening. According to conspiracy theories, the plot describes what is really happening. Instead, the conspirators (and their complicit or unwitting allies in the media) feed us the cover story as the "official" story to keep us in the dark. 'Lee Harvey Oswald acted alone' is one cover story. 'The moon landing really happened' is another. The cover story is what you'd think really happened, if you weren't in the know.

Considering these four features gives us a handy way to describe conspiracy theories beginning with the faked moon landing conspiracy.

"The Moon Landing Was Faked" Theory

Agenda. To achieve a public relations victory over the Soviet Union during the cold war.

Conspiratorial Group. NASA and several "cold warriors" at the highest level of government.

The Plot. Ideally, the conspirators would have preferred to actually send someone to the moon, but there were too many technical difficulties. The conspirators had to adopt plan B, which was to harness the immense power of Hollywood to stage the moon landing here on earth.

Conspirators' Cover Story. By marshalling massive amounts of expertise, we were able to overcome the technical difficulties. We really landed on the moon on July 16, 1969, mere months prior to the deadline set by JFK.

Big Oil Cover-up or Chinese Hoax?

Is the idea of global warming the result of a Chinese Hoax aimed at torpedoing US industrial might? Or are we only barely seeing the scope of global climate change because of the nefarious and cash fueled efforts of Big Oil? The debate over global climate change has spawned two separate conspiracy theories. Looking at these two conspiracy theories can help us get a handle on how one might judge the worth of conspiracy theories. Not all conspiracies are equally good. Put another way, there are ways in which we can determine which conspiracy theory would "win in a fight."

On November 6th 2012, Donald J. Trump, who at the time was a private citizen, tweeted, "The concept of global warming was created by and for the Chinese in order to make US manufacturing non-competitive" <https://twitter.com/realdonaldtrump/status/265895292191248385>. He has since, prior to the run-up to the 2016 election, walked those claims back. Still he insists that the story provided by climate scientists is not factual because the scientists "have a very big political agenda" (interview by Leslie Stahl, *60 Minutes*, October 15th 2018).

As far as conspiracy theories go, this one is a little half-baked. In the earlier version, the agenda is clear, but the conspiratorial group only mentions the Chinese government. It's not credible to assume that Beijing can dictate what climate scientists write in their reports. After all, most of the climate scientists are Western scientists who are not under the direct influence of the Chinese government. The later version identifies the scientists as the conspiratorial group. Unlike the first version of the conspiracy theory, the conspiratorial group more plausibly has the means to influence climate reports. The problem for the latter version is that it is not clear what the "very big political agenda" is. Still we can synthesize a well-designed conspiracy theory from these two versions.

"Climate Change Is a Chinese Hoax" Theory

Agenda. The government of the People's Republic of China wants to displace the United States as the world's foremost economic power.

Conspiratorial Group. The Chinese government allied with world's climate scientists. Climate scientists in this case are either unwitting agents of the Chinese government or allies of convenience due to a "very big political agenda" with anti-capitalist sympathies.

The Plot. Using fossil fuels is a cheap and easy way to grow a country's industrial might, as long as demand remains low. Given that the US and the rest of the West is so far ahead of China, Beijing decided that limiting how much the US and the West use fossil fuels will help them achieve their goals. Limiting the West's consumption of fossil fuels would reduce demand (making fossil fuels even cheaper for China) and hobble the West's industrial growth. To make that happen, China enlists the intentional or unwitting aid of Marxist and anti-capitalist climate scientists to concoct a story about climate change.

Conspirators' Cover Story. Climate scientists are apolitical and anthropogenic climate change is really happening, and only the uneducated believe otherwise.

On the other side of things there is another conspiracy theory related to climate change. According to this theory, the major oil companies are aware of the hazards of the continued use of fossil fuel. Wanting to maintain high profits and make the most out of their investments, the oil companies sow doubt about the scientific evidence in an effort to prevent or delay regulations harmful to the industry. The nuts and bolts of this conspiracy theory are detailed in *Merchants of Doubt* by Naomi Oreskes and Erik Conway.

"Big Oil Cover-up" Theory

Agenda. To maintain profits and not leave any oil money in the ground.

Conspiratorial Group. Oil company executives and a handful of scientists acting as "merchants of doubt."

The Plot. As detailed in *Merchants of Doubt*, oil industry groups have funded a few scientists to vigorously and loudly question the completeness of climate science. The goal is not to dominate or change climate science, but to create enough doubt to assert that the science is not settled. This provides cover for politicians and regulators to delay the serious regulation of carbon allowing oil companies to make the most of their investments.

Conspirator's Cover Story. Climate change is a hoax, and scientists are merely pursuing their political agendas.

Notice that the cover story of each of these conspiracy theories actually ends up as the plot of the other conspiracy theory. What we have here is a cage fight between the two theories. Both can't be true. One says that anthropogenic climate change is happening and the conspirators are trying to make it look like it isn't. The other says the exact opposite. Not only is this a cage match, but it is an ideal cage match for evaluating conspiracy theories.

Is It All Political?

In the US, our society is probably more polarized than it has been since the Civil War. We (John and Michael) remember a time when liberals and conservatives would get their news from the same news source. Those days are gone. Instead, if you are right-leaning, you tend to get your news from right-leaning sources, and if you are left-leaning, you get your news from left-leaning sources.

Everyone tends to buy in to a meta-conspiracy theory. If you're right-leaning and a news story conflicts with what your particular tribe believes, then you might say, "That's just what the liberals want you to believe." If, on the other hand, you are left-leaning, then you might decry uncomfortable information as "fake news." Due to a phenomenon called political sorting, where someone tends to consult the party's platform to determine what they believe, climate change has been wrapped up in this larger meta-conspiracy. So, if you're left-leaning, then you tend to think that the big oil cover-up is more plausible, and if you're right leaning you tend to think the Chinese hoax is more plausible.

The problem is that the way you tend to vote in elections has no bearing on which theory is actually true. So, neither side can trust the intuitions of what their particular tribe believes. Truth and falsity are neither liberal nor conservative. Neither Democrats nor Republicans (nor any other political party) have a monopoly on the truth. What makes this cage match even uglier is the fact that both sides claim that they have the weight of science behind their theory. Of course, someone could resolve that conflict if they actually did the science, but that takes years of training, and few are willing to take those steps. So, we have to figure out another way to resolve the match.

Let's Treat Every Theory as a Scientific Theory

Philosophers of science spend a lot of time trying to identify what makes one scientific theory better than another. By attending to the history of science and the practice of current scientists, we are able to draw out the forms of reasoning that scientists use, or ought to use, to endorse one theory over another.

There are exciting, famous examples, such as the paradigm shift from an Earth-centered universe to a sun-centered solar system, and more mundane examples, such as the reconstruction of sauropod necks as horizontal, rather than vertical. From the rich, varied, and interesting practice of scientific theory selection, we can get a detailed picture of the scientific reasoning that underwrites the selection of scientific theories.

If we treat conspiracy theories seriously as *theories*, then we can use the tools that philosophers of science have extracted from looking at science to help us judge which is best. This toolkit is quite large, but we think a few interesting tools will serve to determine which of the Chinese Hoax or Big Oil Coverup will win in a fight. Even better, the couple of tools we use here could be used to determine if other conspiracy theories have what it takes!

How to Abuse Sherlock Holmes with Conspiracy Theories

In Sir Arthur Conan Doyle's books, Sherlock Holmes routinely deals with conspiracies. In one particular story (*The Sign of the Four*), Sherlock Holmes declares his dictum, "Once you eliminate the impossible, whatever remains, no matter how improbable, must be the truth."

Holmes's strategy, put this way, reveals a couple of the interesting features of reasoning about theories. First, it overstates the power of eliminating the impossible. You shouldn't assume that just because you've shown that something is possible that it must be true. It is possible that we have mailed you a winning ticket to the Powerball lottery, but we haven't. (Sorry.)

At the heart of Holmes's method is eliminating the impossible. This is a part of good reasoning about theories. Good theories are the ones that we try hard to demonstrate are false using the available evidence, but despite our best efforts we still cannot falsify. When we find observations in conflict with our favorite theory, we should reject our theory. Rejecting even our most cherished theories is part of what makes scientific theories objective.

Our choice of theory is not open to mere personal preference, rather, the observations we make tell us whether or not we get to continue to believe our theories. Or put another way, if we design our theories and experiments well enough, the world itself (rather than our culture, our friends, or trusted authorities, or our political persuasion), will tell us what we get to continue to believe.

Holmes determines that Moriarty is engaged in a conspiracy to commit the crime in question not because Holmes wants Moriarty to be guilty, but because Moriarty *is* guilty. In eliminating alternative theories by the careful examination of the facts, Holmes zeros in on the truth. This is, according to Karl Popper, the strategy scientists ought to deploy in zeroing in on the true scientific theories—that the sun is the center of solar system and that sauropods did not raise their necks to graze the tops of the colossal, mesozoic trees.

All is not puppies and rainbows with falsification, however. One problem with falsification as a tool is that observations do not directly show that theories are false. The real world is messier than most works of fiction. Theories are connected to the observations in conjunction with other theories. This connection makes it possible to play a kind of shell game as to whether it is the theory or one of the assumptions we need to make in order to test the theory that is falsified when we get new evidence. This was a central finding of two philosophers of science, working independently, Pierre Duhem and Willard Quine.

One important observation in falsifying the theory that sauropods had vertical, craning necks concerns the blood pressure it would take to circulate blood up such a huge neck while grazing. Computer models suggest it would take a ridiculous amount of energy just to raise such a prodigious crane, and, as a result, it is "impossible" that the sauropod body could maintain a healthy blood pressure with its neck craned. As a result, we reject the vertical, crane-neck theory in favor of a horizontal neck for sauropod morphology. If you were a fan of the vertical theory, you could respond that the blood pressure observation only makes the crane neck theory false if we assume that sauropods had similar vascular systems to ours. So, with a little imagination you could instead assume that sauropods had dozens of hearts and a monstrous energy production system, saving your favorite theory. Duhem and Quine saw that this was possible in scientific investigation and foresaw the problems it might cause. This problem for falsification is often called Duhem-Quine underdetermination, since the observations *do not determine* whether a contrary observation falsifies the theory or one of the assumptions needed to deduce a testable consequence.

Underdetermination is a problem for falsification as a tool in theory selection, but this kind of underdetermination highlights an interesting feature of theories in general, and conspiracy theories in particular. Conspiracy theories that can weather some initial inspection exploit the fact that they add a bunch of additional assumptions that can take the blame when something goes wrong. Conspiracy theories are notoriously difficult to falsify because they have more moving parts than standard scientific theories.

The cover story, for one, allows for a conspiracy theory to do just as well on the evidence as any non-conspiracy theory. The belief that we landed on the moon is not falsified by any of the facts, but neither is the conspiracy theory because of the success of the conspirators in achieving their agenda with a well-conceived plot and good cover story. The agenda, the conspirators, and the plot are all additional assumptions that could be blamed or altered in light of some new fact that would falsify the conspiracy theory.

Imagine we discover that *all* of the Hollywood sound stages were being used for other purposes during the theorized time for Plan B—the faking of the moon landing. The conspiracy theorist is free to suggest that Hollywood was not where the moon landing was faked, but that a sound stage was built in, say Winnipeg, to do so in even greater secrecy.

A conspiracy theorist has tools to abuse Holmes's dictum. The conspiracy theory makes what seemed impossible actually possible, preventing the un-tinfoiled from zeroing in on the truth. The good news for conspiracy theorists is that well-designed conspiracy theories can't be proven false. The bad news is that is not enough to show that the conspiracy theory is plausible. Put another way, "Holmes! Eliminating the impossible is not as easy as you thought."

Predictions, Not Clever Construction, Put Theories to the Test

We think that clever constructions make for good conspiracy theories in fiction, but more than clever construction is needed to make a conspiracy theory plausible in real life. In light of Duhem-Quine underdetermination, philosophers of science have identified a different way of understanding Holmes's dictum for accepting and rejecting theories. It's possible to evaluate the various parts of a theory on the basis of the predictions they do make.

Conspiracy and non-conspiracy theories of the moon landing make the same prediction about whether or not we will observe a "moon landing during the month of July 1969." William Whewell calls these predictions, *predictions of the same kind.* After that, the two theories radically differ in their predictions. The various parts of the plot and the agenda of a conspiracy theory make their own predictions. In a good theory, we should be able to check the various assumptions as well as the theory in question by looking to see if the predictions they make are true. This what Whewell would call *predictions of a different kind.*

One advantage of this strategy over the falsification strategy, is that in a good theory we should be able to gather evidence for the assumptions and the theory using the predictions they make in other circumstances. In the case of the sauropod neck, we cannot check directly whether these dinosaurs had a very different cardiovascular systems, but we have independent reason (evolutionary theory) to avoid dumping the falsification on this assumption, rather than the crane-neck hypothesis. Checking the predictions made by the various assumptions takes away the shell game, making it harder to hide the problematic observations on certain assumptions, rather than the theory in question.

Another interesting thing about prediction is that it permits us to decide when a theory is *making predictions,* rather than merely accommodating the observations. For any set of observations, it's possible to modify your favorite theory to makes sense of them. Prediction is much more difficult, since it requires that you get new evidence correct as well. This is true even of theories that are about the past; as we investigate what happened, we uncover new observations that we want our theory to get right without having to alter the theory or assumptions to get it right.

If we always change our theory to fit with new observations, we are building theories that will be bad predictors, and we are not in fact, predicting anything when we continue to alter the theory or assumptions in the face of new evidence. In effect, we want our best conspiracy theories to make predictions (of a different kind), but not just of the results (of the same kind) they were built to get right.

Conspiracy theories, unlike most scientific theories, face two issues on the predictive front, in part because of how they are constructed. First, since conspiracy theories have lots of moving parts, they require a lot of extra assumptions to be true, for example the plot requires a lot of work, sometimes by a large number of people as well as massive deployments of wealth or power. These assumptions make predictions that we could check. Yet, conspiracy theorists usually don't do the work to check whether the predictions of these assumptions work. Even worse, if they do and their predictions (of a different kind) fail, they use Duhem-Quine underdetermination to shell game the problem. The conspiracy theorist could say that the massive deployment of wealth necessary could be untraceable. After all nefarious, badfolk have truly gifted accountants!

Relative to their scientific counterparts, conspiracy theories are more open to checking whether they are true, because they build in additional checkable content by way of the plot. On the other hand, relative to their scientific counterparts, conspiracy theories produce fewer successful predictions, in part because failures of prediction can be shell gamed in some cases.

Second, some of the assumptions of the conspiracy are notoriously difficult to check in terms of their predictions. The cover story is a prime example of this. The cover story, according to conspiracy theories is supposed to describe the world the way it *seems* and not the way it *is*. The predictions, in terms of observations, that the cover story makes are exactly the same predictions that would be made if the conspiracy theory were false. After all, that is what they want you to think! So, if the cover story makes a prediction that we should see news reports that support the official story, does that prediction confirm the conspiracy theory over the official story? Presumably not. We'd call that prediction evidentially neutral between the two theories.

When we compare scientific theories, we ask whether our theory is more predictively successful than its competitors. We can measure this in two ways. We can count the number of successful predictions and we can look at the extent to which those predictions make the theory more plausible than its competitors. Ironically, the scientific theories that tend to work the best are those theories that optimize the predictive power of their theory with the complexity of the theory. A simpler theory that makes better predictions is always preferred to a more complex theory that has to be constantly modified due to failed prediction.

Here again, we can see an uphill battle for conspiracy theories relative to scientific or common-sense theories. In many cases, the work to test a conspiracy theory will take a bit more than modifying the theory from our barstool. Of course, our goal isn't to determine whether or not conspiracy theories are the best of all theories, merely that the philosophy of science toolkit can help determine which conspiracy theory will win in a fight.

The Chinese Hoax versus Big Oil—the Final Throw Down

Both the Chinese Hoax and the Big Oil Conspiracy make predictions. They provide a cover story, which makes checkable predictions and accommodates the facts as they appear to be. They offer a plot, with components that make checkable predictions. And, they offer us a set of conspirators, and this also offers us checkable predictions. So, if we hold these theories to the standard that they are not open to simply changing the nature of the cover story, the plot, or the list of conspirators, then we can use their predictions to see which is the better conspiracy theory.

The Chinese Hoax offers the Chinese, and their allies in climate science, as the conspirators and the use and regulation of fossil fuels as the core of the plot. Sure, it isn't very specific, but if the Chinese had created the concept of climate change in order to ruin the American economy by undermining the use of fossil fuels, we would expect to find them disregarding or ignoring the regulation of fossil fuels in China. This conspiracy theory fails on this sort of prediction. China is in no way dodging its responsibility in fossil fuel regulation. In fact, China is aggressively pursuing alternative energy programs. Without changing the theory, it is very difficult to see why they would do so. Additionally, it would be difficult to account for why apparently independent groups of scientists from non-US and non-Chinese countries are spending their research time and effort on investigating and defending the idea that anthropogenic climate change is happening. These predictions seem to fail.

Furthermore, while the theory itself is silent on the exact motive behind the behavior of scientists, it seems that the theory does predict that climate change should not be detectable since it isn't real. These scientists are in effect demonstrating that a key part of theory is getting the predictions wrong. Again, a failure of predictive success for the Chinese Hoax theory.

On the other hand, if scientists are in on the game (as one version of the hoax seems to suggest), then we should expect to find evidence of collusion, probably in the form of wealth exchange. Again, a predictive failure, unless these scientists are very good at hiding a massive influx of wealth. We've yet to meet a climate scientist whom we would say was living off Chinese money. Of course, they could be colluding on the basis of expected future wealth. This suggests that we might predict the way they would save for retirement would differ from other academics among other things; again this prediction seems to fail.

The Chinese Hoax has components that make predictions. Some of these predictions fail. The Chinese Hoax can avoid the shell game by checking the predictions the theory makes and the independent predictions its assumptions (the plot, the conspirators, the agenda, and the cover story) make. Ultimately, the problem for the Chinese Hoax is not merely its failed predictions, but that none of the theorists bother to check the many predictions it does make. The remaining issue with the Chinese Hoax, understood as a theory is whether or not its main theorists continue to alter the theory to accommodate (rather than predict) new observations. Alas, there aren't many theorists working on this conspiracy theory, so it is difficult to tell whether or not they would engage in this sort of chicanery. Our prognosis: for the Chinese Hoax, you can probably take off your tinfoil hat. But, let's see how the Big Oil Conspiracy fairs on these basic tests of a theory, before we make our final decree.

Let's take a look at a few of the predictions (of a different kind) that the Big Oil Conspiracy makes. In some cases, the Big Oil conspiracy theory makes predictions that are contrary to the Chinese Hoax theory. Given that it holds that anthropogenic climate change is happening, it predicts that many countries pursuant to their own climate investigations will be investing in carbon neutral energy sources. The Chinese Hoax states that some countries, notably China, won't. The facts are that China is aggressively investing in alternative energy sources. So, score one for the Big Oil Cover-up over the Chinese Hoax.

In the same vein, the Big Oil Cover-up theory predicts that scientists are making their money by doing science, and not by getting that sweet bribery cash from the Chinese. Once again, it seems that the Big Oil Cover-up got this prediction right, while the Chinese Hoax didn't. Just looking at those two predictions, the score is 2–0.

The Big Oil conspiracy theory also makes some more predictions. According to the plot, agenda, and what we know about business, large energy companies are loath to give up their share of the energy market. Add that insight to the cover story which says that Big Oil knows that anthropogenic climate change is happening and at some point that reality will affect consumer preferences. So, the Big Oil conspiracy suggests that large oil companies will be making investments into alternative energy programs to maintain market share in the future. This is exactly what is happening, if we can believe their ads. The Chinese Hoax conspiracy is silent on this one, without modification.

The Big Oil Conspiracy theory also predicts that oil companies will be aware of some of the consequences of global warming, like rising sea levels. Since oil extraction often occurs in the ocean, this means that oil companies will be affected by this consequence of global warming. Since according to their agenda, oil companies want to protect their investments, the big oil conspiracy predicts that oil companies will have made plans to adapt their operations to deal with rising sea levels. That is just what they did, according to a report by Lieberman and Rust! If the Chinese Hoax were true, we might expect the opposite.

These are just a handful of the cases where the Big Oil Cover-up seems to get it right and the Chinese Hoax gets it wrong. Does this show that the Big Oil Cover-up is true? Well, it does seem more credible than the Chinese Hoax at this stage, but it may be that both theories are strictly speaking false, even if they both can't be true. However, the Big Oil theory is better supported by the evidence, at this stage.

Does it mean that the Chinese Hoax is false? We haven't proven that it is. As we've seen, that is easier said than done. What we have shown is that one theory makes better predictions. If we're rational, we might prefer a theory that makes better predictions to one that doesn't. After all, we may have to place wagers, in the way of preparation, on one of these theories, and if we place our wagers on the theory that makes better predictions, then we are more likely to get a better return on our investments. From that perspective, it's clear that when these two theories go toe-to-toe, the Big Oil theory with its better predictions is the one that wins the fight.

18

The Wrong Thinking in Conspiracy Theories

BRENDAN SHEA

> In our reasonings concerning matters of fact, there are all imaginable degrees of assurance, from the highest certainty to the lowest species of moral evidence. A wise man, therefore, proportions his belief to the evidence.
>
> —DAVID HUME, *An Enquiry concerning Human Understanding*, Section 10

> The confidence that individuals have in their beliefs depends mostly on the quality of the story they can tell about what they see, even if they see little.
>
> —DANIEL KAHNEMAN, *Thinking, Fast and Slow*, p. 88

Conspiracy theories vary widely in their content, the individuals and groups who believe in them, and in their effects on the behavior of these believers. For this reason, it may be difficult or impossible to come up with a completely general definition of *conspiracy theory* that captures all and only those theories that fit under this general label. Nevertheless, there are a significant number of conspiracy theories that share something like the following form:

> There exists a certain small group of people who share a certain characteristic such as race, religion, occupation, or nationality. They have secretly undertaken actions that have harmed, or are intended to harm, me and people like me. The fact that these actions have not generally been recognized is due to the conspirators' ability to conceal evidence of this.

Within the general scheme, there is plenty of room for variation. For example, the conspirators may be anonymous figures living otherwise unremarkable lives, or they may be well-known and powerful political, religious, or media elites. Similarly, some purported conspirators actively wish harm upon the believer and others—such as conspiracies positing "traitors" or "spies" working to ensure their own country loses some conflict—while others are held to have much more mundane motives, such as the desire for money or power. In this latter case, the harm in question may simply be an especially unpleasant side effect, though one that was foreseen by the conspirators.

Finally, the harms attributed to the conspirators' actions come in a number of forms. So, for example, it may be that the actions of the conspirators have led (or will lead) to the deaths of particular individuals, financial crises or crashes, military defeats, outbreaks of disease or illness, the overthrowal of the government, and so on.

Conspiracy theories of this type all crucially involve failures of what philosophers often call *inductive reasoning*, which involves using our available evidence to determine what is probable or likely to be true. Inductive reasoning is usually contrasted with *deductive reasoning,* which involves attempts to *prove* with one-hundred-percent certainty that a conclusion follows. As it turns out, inductive reasoning makes up a huge part of our day-to-day lives. We reason inductively, for example, when we try to determine what was the *cause* of some event that we just observed, or when we try to figure out what the *effects* of this same event might be. We also reason inductively any time we make predictions about the future, or decide whether to trust what we've read or heard, or make generalizations about a large population based on the smaller sample that we're familiar with.

For this reason, conspiracy theories, and the errors of inductive reasoning that they exemplify, should be of interest to all of us. After all, if it turns out that many of the crucial errors committed by conspiracy theorists are ones that we ourselves are prone to, this will provide a strong reason for thinking hard about our own beliefs, and the process by which we have arrived at them.

Don't Believe Everything You're Told

Conspiracy theories often serve as simple, attractive rivals to other, more complex theories about politics, history, or science. So, for example, where political scientists may offer theories that tie the outcome of a particular election to factors such as economic conditions, demographic shifts, incumbency bias, and the relative appeal of the candidates' platforms and personae, conspiracy theorists often see the hidden hand of conspirators as being responsible for unwelcome outcomes. Similarly, where mainstream medical and scientific research suggests that conditions such as autism, drug addiction, or obesity have complex causal backgrounds, conspiracy theorists might reply that these bad things are actually due to the hidden side effects of vaccines, the clandestine activities of the CIA, or the machinations of "Big Ag."

One way in which conspiracy theories are distinguished from their mainstream rivals is their method of origin and spread, which is often outside traditional scientific and academic channels. In the modern era, for example, conspiracy theories often begin in the so-called "dark corners" of the Internet, as opposed to in peer-reviewed journal articles. They then spread, via both alternative media sources and social media, to larger and larger audiences. To what extent should this sort of difference in origin matter to the credibility of the theories in question?

The Scottish philosopher David Hume (1711–1776) takes up a very similar question in the "Of Miracles" section of his *Enquiry concerning Human Understanding*. Hume was among the first to clearly distinguish between inductive and deductive reasoning, and his account of the problems inherent in inductive reasoning has influenced (and often troubled) scholars studying inductive reasoning ever since. In "Of Miracles," Hume considers whether or not we should ever believe peoples' accounts of miracles. His answer is a resounding "No!" and many of the reasons he provides are applicable to conspiracy theories as well.

Hume recognizes that the reasons people believe in miracles—because they hear or read about them from sources that they normally trust—are based in the same sort of inductive inference that underpins many of the things we believe. For example, nearly all of our beliefs about history, scientific theories, current events, and even the lives of our closest friends and family are, of necessity, based on what textbooks, teachers, newspapers, and other people tell us about these things. Because of the probabilistic nature of inductive inference, this means that is always *possible* that these sources are incorrect. However, we don't normally take this possibility as grounds for dismissing everything we hear or read. So, what makes reports of miracles (or conspiracy theories) any different?

Hume provides a number of considerations for treating reports of miracles differently than other sorts of "testimony," many of which are applicable to conspiracy theories. First, the chain of testimony supporting miracles often looks quite different than that of ordinary events. Miracles are almost universally said to have occurred long ago and/or in places far away, and under conditions that would have made it difficult or impossible for any skeptic to check on the truth of the claim. In conspiracy theories, by comparison, it is often held that the conspiracy theory is happening "right now!" or "under our noses!" However, just as in the miracle case, it's a central part of the theory that there can be no possible recording or confirmation of the conspiracy, since the conspirators have prevented this (perhaps by murdering witnesses or manipulating the media). The fact that reports of miracles and conspiracy theories haven't been and can't be, checked out by skeptical listeners doesn't mean that they are necessarily false, of course. What it does mean, however, is that these reports lack the sort of safeguard that comes with most testimony regarding strange or unlikely events—that is, if they *were* false, we would likely have some evidence of this.

A second key difference Hume notes relates to the *motivations* of those who talk about miracles. After all, one reason that miracles matter so much is that they can serve as evidence for the truth of certain religious views. This provides a strong motivation for people who already hold these religious views to believe in such reports (after all, we all like being shown right!), and it *also* provides motivation for them to spread these tales, even if they don't fully believe in them. After all, telling tales of miracles might win converts for the faith, or signal to other members of the group your "loyalty to the cause." Something quite similar can be said of many conspiracy theories—insofar as belief in these theories is closely linked to membership in some group, we have good reason to doubt the impartiality of those telling tales of conspiracies.

Finally, Hume observes that, while one might think that the sheer strangeness and outlandishness of miracles would make people less likely to believe and repeat them, experience shows that sometimes the opposite often seems to be the case—people seem to *enjoy* believing and repeating stories about events that are utterly unlike things they have experienced themselves. This, again, has close analogues with conspiracy theories. Odd as it may seem, the very claims of a conspiracy theory that seem the furthest detached from evidence and ordinary experience may be the claims that encourage its spread.

Making Mistakes

In the generations since Hume first wrote, scholars in disciplines ranging from philosophy to economics to statistics to psychology have studied the nature of inductive reasoning from a variety of perspectives. While many of these investigations have aimed at uncovering better methods for inductive reasoning, others have aimed at figuring out how good ordinary humans are at inductive reasoning in a variety of contexts. Most of us do well enough when the conclusions of inductive reasoning concern our immediate experience, for example—we learn quickly to avoid hot stoves, or to avoid drinking bottles labeled "poison," but it is much less clear how successful we are when it comes to dealing with big-picture issues regarding statistical or causal reasoning in areas such as economics, science, or politics. These, of course, are precisely the areas where conspiracy theorists are most prone to get things wrong. So, why might this be? And just how common are these errors?

Starting in the late 1960s, two Israeli psychologists—Amos Tversky and Daniel Kahneman—began investigating just these sorts of questions. In a series of influential articles, they argued that humans are not intuitively "good statisticians," and they make a number of *systematic* mistakes when engaging in inductive reasoning. Tversky and Kahneman's research has had an impact far behind psychology, and in particular caused considerable problems for the view (once common in both economics and some areas of philosophy) that humans generally acted *rationally.* While Kahneman and Tversky don't explicitly consider the problem of belief in conspiracy theories, their work provides a helpful framework for identifying and classifying many of the major inductive mistakes that conspiracy theorists make.

A foundational concept of Kahneman and Tversky's approach is that we make many decisions using intuitive *heuristics,* or simple rules for making inductive decisions. In particular, they suggest that, when we are faced with making a complex decision, we often (without realizing it) "substitute" a simpler, easier-to-answer question, and answer that instead. And while this may work well enough in many day-to-day cases, it can also easily lead to fallacious reasoning of the sort exemplified in conspiracy theories.

The Story Just Fits

Conspiracy theories often begin with the intuition that some bad event—a recession, an outbreak of a disease in the local community, or a school shooting—cannot be adequately explained by any combination of normal causal processes discussed by scientists, public health officials, or psychologists and sociologists. They then conclude that this event must have been caused by a carefully planned process (instigated in secret by the conspirators!) that was designed to result in just this sort of outcome. This way of reasoning exemplifies what Kahneman and Tversky label the *representativeness heuristic,* in which the probability of a certain process P causing event E is judged solely by the "resemblance" between P and E and NOT by any careful consideration of how probable it was that P actually occurred, or the potential alternatives to P, or even how good of evidence for P we happen to have.

In the case of conspiracy theories, the representativeness heuristic might explain several inductive failures. First, it accounts for the way conspiracy theorists often seem to ignore the comparative *base rates* of "bad things caused by a combination of ordinary factors" versus "bad things caused by powerful secret organizations working in secret to cause just this sort of harm in each and every gory detail." While the resemblance heuristic pushes us toward the conspiracy story (since it better "resembles" the bad thing in question), this is a bad inference. After all, the vast, vast majority of the harms that we in incur in life are NOT the result of explicit conspiracies intended to cause this exact outcome, but instead are the result of perfectly mundane causal factors acting in combination (that is, plain old "bad luck").

For similar reasons, the representativeness heuristic can plausibly account for conspiracy theorists' tendency to posit highly specific causes for events that are better explained by appeal to statistics. So, for example, small samples are more variable than large samples, and so we should be very careful in drawing conclusions based on what we have observed in small samples, even if the sample in question seems odd to us. So, for example, if two people in a small office of ten people each have a heart attack during the same month, this might seem unusual, but it doesn't provide strong evidence the office coffee has secretly been poisoned by management seeking to save money on future pensions. By contrast, if 200 people in an office of 1,000 people suffer such attacks in a month (the same percent, but a much larger sample), this really does suggest something out of the ordinary is going on. However, in practice, conspiracy theorists (along with the rest of us) systematically overlook this difference in sample size, and too often jump to conclusions on the basis of small samples.

For similar reasons, the confidence we have in our conclusions about the causes of events ought to reflect the strength and variety of evidence that we have seen—after all, it is surely better to read ten high-quality journal articles and one moderately plausible social media post about a conspiracy theory than just the moderately plausible blog post. However, the representativeness heuristic (which ignores quantity or quality of evidence and cares *only* about its fit with a theory) can lead us to ignore this and, in some cases, to feel *more* confident in our conspiracy theory after reading just the social media post, since there are no additional sources to interfere with the nice clean fit between this story and our believing in the truth of the theory it describes. Basically, once we decide to give the social media post any credence whatsoever—as opposed to simply dismissing it out of hand—it can be very difficult to not *overweight* its value as evidence.

Problems with Probabilities

The decision to adopt a conspiracy theory can be thought of as a sort of "bet" about the way the world will turn out, and what the "winning strategy" for living in such a world will be. So, for example, if I suspect there is a good chance that the members of the US Federal Reserve Board are an evil cabal intent on crashing the world economy to enhance the wealth of their corporate masters, I might buy gold and bury it in my back yard to hedge against this. If I assign a significant probability that pharmaceutical companies have hidden the evidence of vaccines causing autism, I might not vaccinate my children. Finally, if I believe it likely that some suspect group of people is up to no good, I might take action against them, potentially including violence.

Most of us would like to think that we are good at making such bets, since they are crucial to making decisions about how we invest our money, vote, and generally lead our lives. So, for example, it seems obvious that a one percent risk of a bad outcome is different from a five percent chance, which is in turn different from a fifty percent chance or a ninety-five percent chance, and our choices and actions should reflect this difference. Unfortunately, according to Kahneman and Tversky, this is not how we actually make these sorts of decisions. Instead, we get things wrong in a number of ways.

First, we tend to focus not on the relative merits of a set of outcomes, but on how we think of ourselves as having arrived at these outcomes, and whether we view them as "gains" or "losses" from a psychological baseline. As it turns out, we care much more about potential losses than we do about potential gains, and simultaneously don't care as much about the relative size of these gains or losses as we should. Conspiracy theorists offer excellent examples of this.

First, in cases where they weigh large potential benefits from a change versus (much smaller) potential losses, they can be highly risk averse, for example when they reject the large potential benefits of vaccines or GMOs on the grounds that there might be hidden health risks associated with these.

Second, in cases where the conspiracy theorists already feel that they are below some psychological baseline, they can instead become *risk-seeking,* and adopt conspiracy theories that lead to highly risky actions in a last-ditch attempt to put themselves back over the baseline, even though the most probable outcome of such behavior would be to put them even further under this baseline than they already feel themselves to be. So, for example, if the members of a certain group worry they are "losing control of their country" to their political rivals, they might respond by abandoning democratic norms or engaging in violence, even though these actions are, on balance, likely to lead to even greater losses.

Prospect theory also suggests that we systematically underweigh the probabilities of some events while overweighting others. In particular, while we sometimes tend to treat extremely unlikely but possible events as being equal to 0, we quickly *inflate* the probabilities of unlikely events once we begin to treat them as being genuinely possible, no matter how "objectively" unlikely they might be. In the case of conspiracy theories, this might plausibly explain the simultaneous urge to 1. dismiss out-of-hand the possibility that the harms that have occurred to them are due to statistical "chance," and 2. vastly inflate the probability that these harms are caused by the secret actions of conspirators.

Can We Avoid Mistakes when It Counts?

So, what's the take-away from all of this? It might be summarized as follows: conspiracy theorists, like the rest of us, notice bad things happening in the world around them. They (again, like the rest of us) are convinced that there must be a cause for these events. However, when they begin to consider what sort of cause this might be, they are led astray by the resemblance heuristic, which predisposes them towards a causal story (the conspiracy theory) that most closely "resembles" the limited samples they are familiar with, and the limited, biased evidence they have reviewed.

This completely ignores the possibility that the events in question are simply the result of statistical "chance." These errors are compounded by the failure to deal with probabilities and "risky decisions" properly, as described by prospect theory. Conspiracy theorists are often attached to some (perhaps imaginary) baseline about the way things "used to be" or the way "nature intended things," and are willing to take risks to avoid accepting losses from this baseline. Simultaneously, they improperly dismiss the possibility of some unlikely events (such as the sorts of chancy processes that *often* explain strange-looking results in small samples) and they inflate the probability of others (such as the conspiracy theory they've heard so much about on talk radio).

In *Thinking, Fast and Slow*, Kahneman argues there are other heuristics and biases waiting to trip us up, beyond those described here. The *halo effect,* for example, predisposes us to (without any evidence!) assign good qualities to people or things we *already* believe are good in other respects, and bad qualities to those we already dislike or distrust. *Outcome bias,* meanwhile, presents us with a false view of the past, whereby we assume that the things that did happen (for good or bad) were *predictable.* This conveniently allows us to avoid giving credit to decision makers for decisions that turned out well while blaming them for decisions that went wrong.

These sorts of processes plausibly lend fuel to the fire of conspiracy theorists' tendency to blame any and all bad outcomes on the actions of the purported conspirators (who, not coincidentally, tend to belong to groups the theory's proponents already hold in ill regard). Finally, and perhaps most concerning, our intuitive sense of how likely a given outcome is, is strongly affected by the detail in which one has imagined or described this outcome. So, the mere act of talking or reading about a conspiracy theory in detail might well serve to inflate our sense of how probable this sort of really thing is.

All of this generally happens without even thinking, and it can happen to even smart, knowledgeable people, since inductive fallacies don't present themselves as defective means of reasoning. Instead, these processes present themselves as a strong feeling that certain theories or ideas are correct, and invite us to adopt and defend these ideas as our own with all of the intellectual creativity and rigor that we can muster. This suggests that that vulnerability to conspiracy theories may be linked to neither ignorance nor stupidity. Rather, it might be that conspiracy theorists are mentally "lazy" in the ways that many of us are lazy, and it is this laziness that undercuts their ability to make cogent inductive inferences.

Belief in a conspiracy theory allows us to avoid all sorts of uncomfortable thoughts, such as fully grappling with the role of chance in events, or the poverty and bias of the news we consume, or the systematic ways in which our sense of what's possible misleads us about what is actually probable. Conspiracy theories reassure us that the bad guys really are all bad, and that, if we stop them next time, we can assure things will turn out well.

If correct, this suggests that there can be significant value in reflecting on the inductive failures of conspiracy theorists, even for those who feel quite confident that they themselves could never fall into the trap of believing in such a theory. Such confidence, as it turns out, may be a poor guide to our actual vulnerability. However, it may be that we can partially inoculate ourselves against conspiracy theories by paying close attention to the *specific* ways in which they exemplify bad inductive reasoning. This, in turn, might make it at least somewhat easier to catch our own errors, and to become better, more careful inductive reasoners.[1]

---

[1] I'd like to thank Todd Kukla for his helpful comments.

19

How Can I Say This without Sounding Crazy?

PAUL LEWIS

I have an unusually vivid memory from childhood of my first encounter with politics. It was the Summer of 1973, and all the major networks were showing the Watergate hearings on television.

Although I didn't watch the proceedings very much, I could not have failed to notice that this one program alone was running day after day, all day long. Also, there were no commercials, which meant that something must have gone terribly wrong in the world.

At the time, I understood this much: some important people, including the President, had been caught lying, cheating, and stealing. I was seven years old, so I could appreciate the joy of mischief, but these guys did not look like they were enjoying themselves at all. It turns out they were only lying, cheating, and stealing to cover up offenses that were much worse.

I have not forgotten the gravity and tension in that living room, where the television was on all day, where I passed through to get a drink from the kitchen, and where my unblinking parents leaned forward on the couch to measure the full dose of corruption and deceit Americans had been forced to swallow. This is not a memory of politics as some bland discussion topic in the classroom. It was politics that I could feel and taste in the air. It was bitter. My dad had been right all along.

The Truth Must Be True

My dad was a cynical and sophisticated political thinker. If some pundit advised going to war to make the world safe for democracy or cutting taxes and regulations on big corporations to expand the economy, my dad knew it was a lie. He had read General Smedley Butler's *War Is a Racket*, he had read the Powell Memorandum, and he had read Daniel Ellsberg's *The Pentagon Papers*. He was always reading and could cite half a dozen historical examples of red-handed political infamy to harden his case, often punctuating his diatribes with a pithy question or phrase. "The truth must be two things," he used to say, pausing long enough to let you glimpse the riddle crouching behind the statement. "It must be true, and it must be told."

Of the many proverbs in my dad's arsenal, this one is my favorite, at least nowadays. By using a double meaning for the word *must*, it expresses something obvious and necessary to the definition of truth, but only to drop an unexpected moral responsibility onto this definition. First, the truth must be true because if it isn't true, then it isn't the truth at all. Terrific. Second, the truth must be told either because it's better that it be told or because something bad will happen if it isn't. When that second thought enters, then unlike most descriptions of truth this one comes with a warning label: if a truth is true but untold, then the *value* of this truth is unrealized.

Logical purists can insist that the truth is the truth whether it's told or not. Fine. I'll direct my appreciation for this double requirement on truth to cases in which the truth matters, to cases in which there is some real social or political value at stake in telling the truth. In these instances, respecting the truth will require knowing what is true, of course, but also knowing how to tell it. Remember this.

Philosophers think a lot about truth, and when they turn their attention to conspiracy theory—or to conspiracy theory theory—this preoccupation with truth naturally rises to the top of their concerns. The study of truth, knowledge, evidence, doubt, and other related terms is its own major field within philosophy. It's called epistemology, a word that combines the ancient Greek words for 'knowledge' and 'explanation', thereby giving a name to the tough, dirty, and thankless job of explaining what it really means to know something—and also what it means to fall short of knowing.

Because ongoing conspiracies actively resist outside efforts aiming at their discovery, they present interesting and problematic cases in epistemology. Lasting provisions of secrecy and obfuscation can even protect historical conspiracies against full disclosure long after they are lapsed and buried by the passing years. A carbon atom, the planet Mercury, a bolt of lightning—these things make no effort to thwart scientists by shredding evidence or disseminating false data, but conspiracies cannot exist without these (and many other) assaults and traps laid against those who would try to know and understand them. They insulate themselves in silence; cloak themselves with camouflage; and sometimes, where corruption, apathy, or ignorance have become normalized, they can even hide in plain sight.

Conspiracies are also protected against scrutiny by a linguistic bias that smears the consideration of any conspiracy theory with an intellectually dubious or unsavory quality. This default setting is unfair. Gravitational theory, number theory, and personality theory, for example, circulate freely without any similar blemish on their credentials: there is a thing we call gravity, it exists, some people formulate theories about it, and that's respectable. On the other hand, there are things we call conspiracies, they exist, some people formulate theories about them, they are crazy and it's time to break out the tinfoil party hats. This prejudice has occupied too much space in both popular and professional conversations on the subject, as if conspiracies are extremely rare (they are not) because conspiratorial secrecy is so difficult to maintain (it isn't). The repetition of this slanted everyday usage has ensured that the conventional wisdom about conspiracy theory generates plenty of conventional folly about conspiracies themselves.

The Truth Must Be Told

Wherever there are real social and political consequences attached to its disclosure, knowing the true makeup of a conspiracy is only half the work: it remains to be told. How hard can that be? This depends upon the *complexity* of the conspiracy itself and the *tools* available to tell the story of its genesis and operation. There are other factors affecting the task, but I will concentrate on these two.

Because the difficulty of telling the truth about a conspiracy depends on the complexity of the conspiracy and the efficacy of the tools for telling it, there could be a conspiracy so complex that it exceeds the capacity of the tools available to adequately narrate it. This troubling possibility may sound like a paranoid conceit—the conspiracy that can't be described even when it's known—but it is not terminal. When the existing tools are inadequate, new ones can be invented.

Isaac Newton lacked the tools to adequately tell what he knew about the mathematical principles of nature; those tools didn't exist, so he created them. But until the requisite tools are invented there may be truths that are knowable but incommunicable for some time. "A very great deal more truth can become known than can be proven," proclaimed physicist Richard Feynman. If this holds true in the physical sciences, then it must carry some weight in the consideration of conspiracies as well.

This circumstance presents a puzzle: if the mechanics of a conspiracy are too complex to explain, then how could the conspiracy possibly function as a co-ordinated operation in the first place? Don't the conspirators have to understand the plot clearly enough to carry it out? No, they don't. If this sounds strange, it's only because you are imagining one particular type of conspiracy, and you have probably drawn your image from a common but faulty assumption of what a conspiracy is. What is a conspiracy anyway, and how many different types of conspiracy are there?

What Is a Conspiracy?

A conspiracy is a network of actors whose intentions, plans, and efforts require protection by secrecy or disinformation if they are to avoid interruption, since the continued operation of the network, in whole or in part, would be jeopardized by their discovery. This is my working definition. The word *network* is better than something simpler like *group* because it recognizes not only the human actors in a conspiracy but also the nonhuman technical connections that hold them together: satellite phones, encrypted files, copper mines, bank secrecy laws, pipelines, tax codes, drugs, unmarked jet planes, counterfeit money, and so on. The network also includes human drives and dispositions like fear, greed, jealousy, vengeance, and so on.

Together these technical connections and human dispositions make up an environment of opportunities in which conspiracies emerge, grow, and die like organisms. Secrecy and disinformation function as the "cell membrane" that keeps the network intact and operational within the larger field of social, economic, legal, and political forces. With this protection, the conspiracy can adapt and evolve in response to emergent threats and prospects. In this sense, a conspiracy is like a bacterium or a rogue cell hiding within a larger body, and the secrecy that defines its boundary makes it difficult to detect.

Some Conspiracy Types

In common usage the word conspiracy is typically associated with wrongdoing. This assumption is not essential to my definition, which allows for a range of instances, including some that most people would call benign (conspiracy to stage a surprise birthday party), malign (conspiracy to assassinate the president), or ambiguous (conspiracy to steal from the rich and give to the poor, or conspiracy to carry out a psychology experiment whose subjects don't know what's going on).

The malign conspiracy is distinguished from the others by its use of illicit methods to achieve morally or politically reprehensible goals. The last line of defense for malign conspirators, once they are caught, is to admit the conspiracy—at least the parts that are not too damning—while offering a moral justification of its overall purpose. "We did it for the greater good of (fill in the blank)." Psychologists and sociologists even have a colorful name for these kinds of rationalizations, which are called blue lies.

In common usage the word conspiracy is also strongly associated with a centralized plot, planned and executed by a mastermind or by a cabal of schemers. If such a plot were discovered, or if only one key insider were to blow the whistle, the vital information about the conspiracy would be exposed and the whole operation would be in grave peril. This fits the description of some conspiracies, but it isn't generally true for all of them, and that's why the assumption of a centralized plot is also inessential to my definition. In fact, precisely this faulty assumption gives rise to the skeptical claim that the mechanics of a conspiracy cannot possibly be *too* complex, otherwise the conspirators could not possibly carry it out.

I can clarify this point with a diagram depicting the two types of conspiracy I am considering. The first is the conspiracy with centralized planning and control, depicted on the left in the figure on page 109.

Engineer Paul Baran first created this illustration to demonstrate the resilience of communication networks, but it works just as well to classify conspiracies, which are a special type of communication network.

The centralized conspiracy has a ringleader or a small group of masterminds at its center. Information and control radiate from and return towards this center. To expose and disrupt the center is to eliminate the network, leaving only disconnections. This is one particular way to describe its vulnerability, which prosecutors exploit by nabbing someone at the periphery and offering a deal in trade for information pointing towards the center. For this type of conspiracy, it is relatively easy to tell the story of its genesis and operation. The story would go something like this: once there was a kingpin who

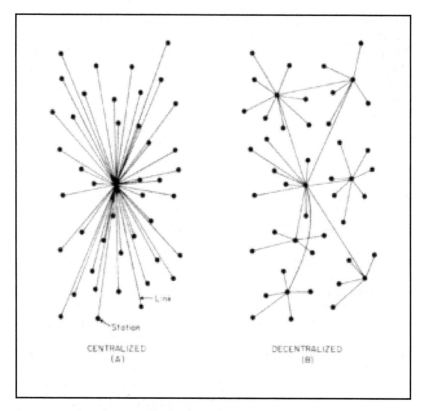

instructed and controlled a bunch of agents who carried out his bidding either in secret or under cover of false pretenses. The end. The most significant feature of this type of conspiracy network is that *the story remains basically the same at any scale of operation.*

The second conspiracy type is decentralized. It is depicted on the right, where there is no absolute center and therefore no ringleader. Instead there are multiple centers joining the flows of information and control, and these centers are themselves joined into a larger network. In this scenario, eliminating a random node or disrupting a single link leaves most of the operations intact and still connected. It is not possible to strike one decisive blow against this type of conspiracy, so in this specific sense it is more resilient than the first type. More importantly, telling the story of its genesis and operations is much more complicated. Most important of all, the story does not remain the same at any scale but *becomes exponentially more complex as the scale of the conspiracy network increases.*

This is why it is not outlandish to imagine a conspiracy network complex enough to exceed the narrative tools required to communicate the truth about it. This is also why the conspirators do not need to know the whole plot in order to carry it out. If an operation such as this could be shown in its entirety, the picture would offer surprise discoveries even to the conspirators themselves. The grand scheme does exist in some sense but not as the activation of a single plan in any ringleader's head. If there is a place for the mastermind, that role goes to the one who stands outside of the conspiracy, sees the whole, and maps it out—the one who connects the dots.

It will require more than words to tell the story. It will require more than a list of characters, motives, plots, reversals, and whatever else is in the narrator's toolkit. Even if you know all the facts of the operation, to tell the story of such a conspiracy within the limitations of conventional narrative is to tell a story that has no center and no main line of development. The story will not make sense. You will interrupt yourself, backtrack, ramble, and your explanation will dissolve into a kind of narrative pandemonium. You will sound like a conspiracy theorist, someone will offer you a tinfoil hat, and at that moment it will occur to you that the following question has become your most immediate concern—how can I possibly say this without sounding crazy? That is what he must have asked himself over and over again.

Death-Defying Acts of Art and Conspiracy

Eventually he figured it out. He taught himself how to say it without sounding crazy, and for this achievement he became one of the first great artists of the twenty-first century—barely. He was found hanged and dead in his Brooklyn apartment on March 22nd 2000, just one day short of his forty-ninth birthday.

In the brief but frenzied three-year period before his death, conceptual artist Mark Lombardi won critical recognition for his painstaking depictions of conspiracy networks in large pencil drawings. He learned to execute these curvilinear schematic diagrams by studying the graphical information techniques—called interlocks—used by forensic accountants and prosecutors to investigate elaborate financial fraud and racketeering schemes. Lombardi's research into the subject matter of his drawings, which he called *narrative structures*, was both careful and extensive, and in some cases he relied upon privileged sources of information. The research was good enough to attract the attention of an FBI agent and a lead investigator into the September 11th attacks, who both found it worth their time to study his drawings for clues and insights.

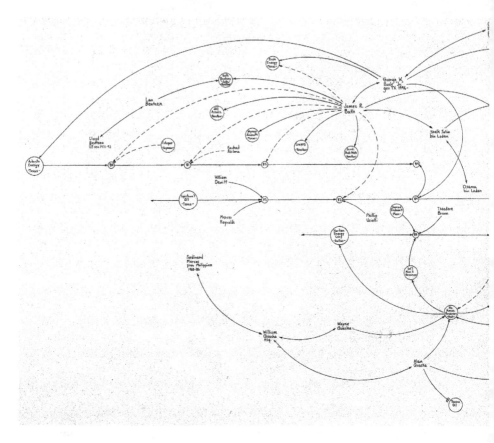

Lombardi studied art at Syracuse University, where he sought the best artistic outlet to express his passion for research. This was the early 1970s. The legitimacy and integrity of the US government was coming unwound as its egregious crimes at home and abroad were becoming public knowledge. Student protestors had been shot and killed at Kent State University. Lombardi joined the Students for a Democratic Society, and he embraced an ethos of socially conscious and politicized art in which the personal experience of the artist is less important than the panorama of social and political relations.

These influences followed Lombardi to Houston, where he worked for several years as a reference librarian at the Houston Public Library. There he pursued his interest in collecting and classifying information; more and more this was information about the major corruption scandals of the day: the war on drugs, financial fraud in savings and loan institutions, and the Iran-Contra affair; more and more Lombardi began to

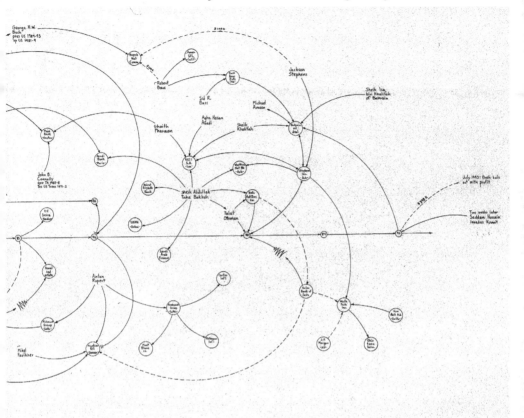

see that these scandals were not unconnected. They were joined into a larger geopolitical network extending decades into the past on a global scale.

The origins of Lombardi's artistic methods can be found in his unrelenting efforts first to understand and then to tell the story of this global and decentralized network of corruption and conspiracy involving government covert operations, money laundering, arms trafficking, bribery, drug cartels, financial fraud, and international terrorism. These efforts brought him to the realization that there was no ultimate separation between the underworld and the upperworld, and if there was a single truth Lombardi was trying to tell, this was it. He must have asked himself: how can I say this without sounding crazy? What he discovered is that this is the kind of truth that can only be told by being *shown*, and then he figured out how to do it. On his business cards he had printed his name and a cryptic reference to his vocation: "Death-defying acts of Art and Conspiracy."

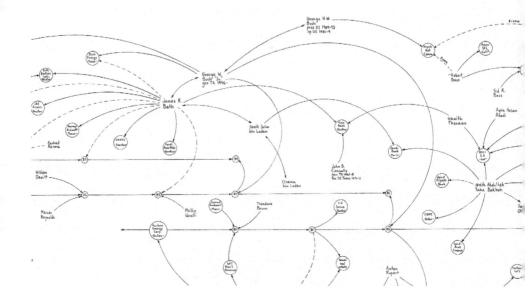

The Mirror of Global Evils

The work most often regarded as Lombardi's masterpiece is *BCCI-ICIC & FAB,* which is currently owned by the Whitney Museum of American Art. The title of the work refers to the subject matter of Lombardi's research into the Bank of Credit and Commerce International, or BCCI. Pakistani financier Agha Hasan Abedi founded BCCI in 1972 with initial capitalization funds from Sheik Zayed bin Sultan al-Nahyan, the Emir of Abu Dhabi, and from Bank of America. Over the next two decades the bank grew rapidly to become the seventh largest private bank in the world, operating in more than seventy countries. By the time it was forced to close, BCCI would be called out, as reported by Beaty and Gwynne, as "the largest corporate criminal enterprise ever."

When journalists began to reveal the scope of its operations, which had persisted for decades, it was natural for many to wonder how such a vast criminal conspiracy could have been kept secret for so long. Here, again, is this stubborn myth in the popular imagination that secrets are difficult to guard. There is also a strong but mistaken expectation that the exposure risk of a conspiracy will rise in proportion to the scope of its operations. While this might be true of centralized conspiracy networks, it is generally false for decentralized conspiracy networks, which are more resilient against defections or breakdowns. BCCI turned out to be the latter.

In 1988 the US Customs Service and the Drug Enforcement Agency conducted Operation C-Chase, an undercover sting in Tampa, Florida, that led to the discovery and successful prosecution of seven BCCI bankers caught laundering drug money for the Medellin cartel in Colombia. Given that this was the first time any international bank had been indicted and convicted for money laundering, it would be reasonable to think that the Tampa bust would be the domino to topple all the rest. This did not happen. To the contrary, field agents were dismayed and frustrated when Operation C-Chase was shut down before it could yield indictments reaching deeper into BCCI activities.

Similarly, John Kerry's senate investigation into BCCI met great resistance from the Justice Department, from the CIA, and from congressional members rallying to the defense of BCCI, which resumed its activities worldwide. Exasperated, the chief investigator on Kerry's committee eventually shared his research with Manhattan District Attorney Robert Morgenthau, who pressed the investigation with renewed energy. This in turn led to a widening spiral of discoveries that shed light on the greater extent of the BCCI network and eventually precipitated its downfall.

Journalists at the time were trying to solve a jigsaw puzzle with a scattering of pieces from trials, hearings, and investigations, and certain questions would not go away. How had this "largest corporate criminal enterprise ever" managed to operate for decades? More specifically, why had certain persons and agencies in the US ignored, delayed, or even opposed the deeper investigation of BCCI? One reporter put the pieces together in the following way:

> The question that continues to haunt the B.C.C.I. story is why B.C.C.I. remained immune for so long. The lengthy investigations of B.C.C.I. provide an answer: B.C.C.I. was a de facto arm of U.S. covert foreign policy in the 1980s, particularly the rough-riding version of that policy embodied by former CIA chief William Casey. B.C.C.I. was deeply involved with the U.S. intelligence community; it was involved in moving money and weapons to Iran and the Nicaraguan contras; it was employed in the covert U.S. effort, co-ordinated through Pakistan, to supply the Afghan rebels with weapons and resources during their ten-year war against Soviet occupation. This accounts for early efforts to block attempts to curb B.C.C.I.; the later cover-up was partly to conceal the government's failure to act upon its early knowledge of the bank's criminality. (Jonathan Beaty, "BCCI: The Trial.")

Mark Lombardi was initially planning to write a book on BCCI, so he was carefully following details of the scandal as they were made public. His research methods incorporated 3x5-inch index cards, and he had accumulated nearly 14,000 of them. BCCI was complex, bankrolling Saudi finance provisions for the development of Pakistani nuclear weapons, money laundering for drug cartels worldwide, illegal weapons trafficking throughout the Middle East, CIA funding of guerilla terrorism in Central America and in Afghanistan, bribery of government officials in South America and in Africa, and much more.

As the scope and complexity of this enterprise revealed itself to be a truly global network, no matter how exhaustive and accurate his research might have been, Lombardi must have felt the inadequacy of his narrative tools. He needed something else. Identifying himself as an investigative journalist, he made a phone call to a Los Angeles attorney, Leonard Gumport, who had worked on a bankruptcy case involving a subsidiary company owned by Adnan Khashoggi, a key figure in both Iran-Contra and BCCI.

As Gumport explained over the phone how to construct an interlock, Lombardi sketched ideas on a napkin and seized on an unexpected insight. Driven by his double passion for art and for investigation, Lombardi had been planning to use the diagrams to organize his research for the book on BCCI, but his conversation with the LA lawyer prompted a realization that the diagrams must become his own original artform. This epiphany was the answer to the vexatious question: how can I say this without sounding crazy?

Although the specific focus of his research was BCCI, Lombardi discovered that as vast as this network may have been, it was still just one node in an extensive network of countless other licit and illicit operations stretching across decades. The CIA, the NSC, and the DIA had all utilized BCCI extensively, for example, and the CIA may well have been involved in its formation from the beginning. State secrecy doctrine and practice will keep critical information about these connections in the dark indefinitely, but enough is now known to validate the general insight Lombardi expressed in his drawings.

BCCI was both a bounded network, which prosecutors and litigators were chasing, and an unbounded network of networks, which almost no one could see. Lombardi understood this and treated BCCI as a corruption scandal of its own time and place, which it was, and also as an allegory for the geopolitical forces that give rise to conspiracy networks over and over again. For Lombardi, BCCI names both a specific bank scandal and the criminogenic field of forces that precedes it and persists afterwards. "Lombardi's diagrams give us a sense not of the network per se but of networking—the processes and operations that join and separate disparate people and events along temporal horizons," writes philosopher Anna Munster.

Lombardi's *BCCI-ICIC & FAB* is both a panorama and an allegory. It was this allegorical quality that prompted criminologist Nikos Passas to characterize the BCCI affair as "the mirror of global evils." Using the thousands of index cards he compiled Lombardi, the artist, pieced together the innumerable fragments of this mirror; he looked into it and made its invisible allegory visible.

This was and will be his most death-defying act.

Bibliography

Abramson, Seth. 2019. *Proof of Conspiracy: How Trump's International Collusion Is Threatening American Democracy.* St. Martin's.

Achen, Christopher, and Larry Bartels. 2016. *Democracy for Realists: Why Elections Do Not Produce Responsive Government.* Princeton University Press.

Allen, Gary, and Larry Abraham. 1976. *None Dare Call It Conspiracy.* Buccaneer.

Anderson, Dave. 1993. Sports of the Times; Jordan's Atlantic City Caper. *New York Times.*

Anonymous. 2009. *The Protocols of the Elders of Zion.* IAP.

Arp, Robert, Steven Barbone, and Michael Bruce, eds. 2018. *Bad Arguments: 100 of the Most Important Fallacies in Western Philosophy.* Wiley.

Baier, Annette. 1986. *Trust and Antitrust.* 1986. University of Chicago Press.

Baker, Emerson W. 2007. *The Devil of Great Island: Witchcraft and Conflict in Early New England.* Palgrave MacMillan.

Barkun, Michael. 2003. *A Culture of Conspiracy: Apocalyptic Visions in Contemporary America.* University of California Press.

Beaty, Jonathan, and S.C. Gwynne. 1993. *The Outlaw Bank: A Wild Ride into the Secret Heart of BCCI.* Random House.

Bell, Daniel. 2017 [1955]. *The Radical Right.* Routledge.

Benestad, Rasmus E. 2016. Learning from mistakes in Climate Research. *Theoretical and Applied Climatology* 126:3–4 (November).

Berkow, Ira. 1994. A Humbled Jordan Learns New Truths. *New York Times.*

Bjerg, Ole, and Thomas Presskorn-Thygesen. 2016. Conspiracy Theory: Truth Claim or Language Game? *Theory, Culture, and Society* 34:1 <https://doi.org/10.1177/0263276416657880>.

Brandt, Allan M. 1978. Racism and Research: The Case of the Tuskegee Syphilis Study. *The Hastings Center Report* 8:6.

Butler, Smedley D. 2003 [1935]. *War Is a Racket: The Antiwar Classic by America's Most Decorated Soldier*. Feral House.

Bynum, William. 2008. *The History of Medicine: A Very Short Introduction*. Oxford University Press.

Byrne, Ruth M.J. 2005. *The Rational Imagination: How People Create Alternatives to Reality*. MIT Press.

Carreyrou, John. 2018. *Bad Blood: Secrets and Lies in a Silicon Valley Startup*. Knopf.

Carroll, Robert Todd. 2015. Forer Effect. *The Skeptic's Dictionary*. <http://skepdic.com/forer.html>.

Carter Center. 2018. Diseases Considered as Candidates for Global Eradication by the International Task Force for Disease Eradication. <www.cdc.gov/mumps/vaccination.html>.

Cassam, Quassim. 2016. Vice Epistemology. *The Monist* 99.

CBS News. 2006. Michael Jordan Still Flying High. *CBS News* (August 20th).

Centers for Disease Control and Prevention. 2017. Rubella Vaccination. <www.cdc.gov/rubella/vaccination.html>.

Centers for Disease Control and Prevention. 2018a. Mumps Vaccination. <www.cdc.gov/mumps/vaccination.html>.

Centers for Disease Control and Prevention. 2018b. Measles Vaccination. <www.cdc.gov/measles/vaccination.html>.

Churchill, Ward, and Jim Vander Wall. 2002 [1990]. *The COINTELPRO Papers: Documents from the FBI's Secret Wars Against Dissent in the United States*. South End.

Clark, Daniel J., director. 2018. *Behind the Curve*. Video. Delta-v Productions.

Cohn, Norman. 1998. *Warrant for Genocide: The Myth of the Jewish World Conspiracy and the Protocols of the Elders of Zion*. Serif.

The College of Physicians of Philadelphia. 2019. The History of Vaccines, <www.historyofvaccines.org/timeline/all>.

Conover, Adam. 2017. *Adam Ruins Conspiracy Theories* (Season Two, Episode 12 of *Adam Ruins Everything*). Video.

Cook, John, et al 2013. Quantifying the Consensus on Anthropogenic Global Warming in the Scientific Literature. *Environmental Research Letters* 8.

Copp, David. 1980. Hobbes on Artificial and Collective Actions. *The Philosophical Review* 89:4.

Crutchley, Peter. 2014. How Did Hitler's Scar-faced Henchman Become an Irish Farmer? *BBC News*.

Darwin, Hannah, Nick Neave, and Joni Holmes. 2011. Belief in Conspiracy Theories: The Role of Paranormal Belief, Paranoid Ideation, and Schizotypy. *Personality and Individual Differences* 50.

Del Vicario, Michela, et. al. 2016. The Spreading of Misinformation Online. *Proceedings of the National Academy of Sciences* 113:3 (January 19th).

Dentith, M R.X. 2018. Conspiracy Theories and Philosophy— Bringing the Epistemology of a Freighted Term into the Social Sciences. In Uscinski 2018.

Douglas, Karen, R. Sutton, and A. Cichocka. 2017. The Psychology of Conspiracy Theories. *Current Directions in Psychological Science* 26.

Dunbar, David, and Brad Reagan. 2011 [2006]. *Debunking 9/11 Myths: Why Conspiracy Theories Can't Stand Up to the Facts*. Hearst.

Dunning, Brian. 2007. The Greatest Secret of Nostradamus. *Skeptoid Podcast* #66 (September 18th).

———. 2018. *Conspiracies Declassified: The Skeptoid Guide to the Truth Behind the Theories*. Adams Media.

Eco, Umberto. 2016. On Conspiracies. In Eco, *Chronicles of a Liquid Society*. Houghton Mifflin Harcourt.

Esquinas, Richard. 1993. *Michael and Me: Our Gambling Addiction . . . My Cry for Help*. Agc Pub.

Federici, Silvia. 2018. *Witches, Witch-hunting and Women*. PM Press.

Feynman, Richard. 1965. *The Development of the Space-Time View of Quantum Electrodynamics: The Nobel Lecture, 1965*. The Nobel Foundation.

Foucault, Michel . 2014. *Wrong-Doing, Truth Telling: The Function of Avowal in Justice*. University of Chicago Press.

Franklin, James. 2001. *The Science of Conjecture: Evidence and Probability before Pascal*. Johns Hopkins University Press.

Franks, Bradley, Adrian Bangerter, and Martin W. Bauer. 2013. Conspiracy Theories as Quasi-Religious Mentality: An Integrated Account from Cognitive Science, Social Representations Theory, and Frame Theory. *Frontiers in Psychology* 4.

Franks, Bradley, Adrian Bangerter, Martin W. Bauer, Matthew Hall, and Mark C. Noort. 2017. Beyond 'Monologicality'? Exploring Conspiracist Worldviews. *Frontiers in Psychology* 8.

Gibney, Alex, director. 2019. *The Inventor: Out for Blood in Silicon Valley*. Video. HBO.

Gore, Al. 2006. *An Inconvenient Truth: The Planetary Emergence of Global Warming and What We Can Do about It*. Rodale.

Griffin, David Ray. 2004. *The New Pearl Harbor: Disturbing Questions about the Bush Administration and 9/11*. Interlink.

———. 2005. *The 9/11 Commission Report: Omissions and Distortions*. Interlink

Halperin, Ian, and Max Wallace. 1999. *Who Killed Kurt Cobain? The Mysterious Death of an Icon*. Citadel.

Hansen, James R. 2005. *First Man: The Life of Neil A. Armstrong*. Simon and Schuster.

Haynes, John Earl, Harvey Klehr, and Alexander Vassiliev. 2009. *Spies: The KGB in America*. Yale University Press.

Herbert, Geoff. 2011. Harold Camping says May 21 Was a 'Spiritual' Judgment Day and End of the World Is Still Coming. *Central NY NEWS* (May 24th).

Hoffer, Eric. 2002 [1951]. *The True Believer: Thoughts on the Nature of Mass Movements*. HarperCollins.

Hofstadter, Richard. 1965. *The Paranoid Style in American Politics and Other Essays*. Knopf.

Hume, David. 2011 [1748]. *An Enquiry concerning Human Understanding*. Hackett.

Jensen, T. 2013. Democrats and Republicans Differ on Conspiracy Theory Beliefs <www.publicpolicypolling.com/polls/democrats-and-republicansdiffer-on-conspiracy-theory-beliefs>.

Johnson, David Kyle. 2018. Entries in Arp, Barbone, and Bruce 2018.

Jones, James H. 1993 [1981]. *Bad Blood: The Tuskegee Syphilis Experiment*. The Free Press.

Kahneman, Daniel. 2011. *Thinking, Fast and Slow*. Farrar, Straus, and Giroux.

Kahneman, Daniel, and D. Miller. 1986. Norm Theory: Comparing Reality to Its Alternatives. *Psychological Review* 93.

Kahneman, Daniel, P. Slovic, and A. Tversky, eds. 1982. *Judgment under Uncertainty: Heuristics and Biases*. Cambridge University Press.

Klehr, Harvey, John Earl Haynes, and Kyrill M. Anderson. 1998. *The Soviet World of American Communism*. Yale University Press.

Klehr, Harvey, John Earl Haynes, and Fridrikh Igorevich Firsov. 1995. *The Secret World of American Communism*. Yale University Press.

Knight-Jadczyk, Laura, and Joe Quinn. 2006 [2002]. *9/11: The Ultimate Truth*. Red Pill.

Kropf, Martha and David Kimball. 2013. *Helping America Vote: The Limits of Election Reform*. New York: Routledge.

Lacapria, Kim. 2015. Jade Helm Concludes. Snopes.com (September 15th).

Le Couteur, Penny, and Jay Burreson. 2004 [2003]. *Napoleon's Buttons: 17 Molecules that Changed History*. Penguin Random House.

Leman, Patrick. 2003. Who Shot the President? A Possible Explanation for Conspiracy Theories. *Economist* 20 (March).

Levin, Jack. 1975. *The Functions of Prejudice*. Harper and Row.

Lieberman, Amy, and Susanne Rust. 2015. Big Oil Braced for Global Warming while It Fought Regulations. *Los Angeles Times* (31st December).

Lukić, Petar, Iris Žeželj, and Biljana Stanković. 2019. How (Ir)rational Is It to Believe in Contradictory Conspiracy Theories? *Europe's Journal of Psychology* 15:1.

Madsen, Kreesten Meldgaard, Anders Hviid, and Mogens Vestergaard. 2002. A Population-Based Study of Measles, Mumps, and Rubella Vaccination and Autism. *New England Journal of Medicine*.

Malice, Michael. *The New Right: A Journey to the Fringe of American Politics*. St. Martin's.

Malone, William Scott. 1978. The Secret Life of Jack Ruby. *New York Times*.

Marrs, Jim. 2011. *The Trillion-Dollar Conspiracy: How the New World Order, Man-Made Diseases, and Zombie Banks Are Destroying America*. Morrow.

Martinez, Tomas Eloy. 1997. The Woman Behind the Fantasy. *Time*.

Mather, Cotton. 2002. The Wonders of the Invisible World. In George Lincoln Burr, ed., *Narratives of the New England Witchcraft Cases*. Dover.

McFadden, Robert D. 2013. Harold Camping, Dogged Forecaster of the End of the World, Dies at 92. *New York Times* (December 17th).

McKay, Ian. 2019. Pareidolia and Apophenia Explained. Owlcation (March 31st) <https://owlcation.com/stem/Pareidolia-Explained>.

McKenzie-McHarg, Andrew, and Rolf Fredheim. 2017. Cock-Ups and Slap-Downs: A Quantitative Analysis of Conspiracy Rhetoric in the British Parliament 1916–2015. *Historical Methods: A Journal of Quantitative and Interdisciplinary History* 50:3.

Memmott, Mark. 2011. 'Rapture' Prophet Camping: Did I Say May 21? I Should Have Said October 21. *NPR: The Two-Way* (May 24th).

Menninger, Bonar. 1992. *Mortal Error: The Shot that Killed JFK*. St. Martin's.

Merlan, Anna. 2019. *Republic of Lies: American Conspiracy Theorists and Their Surprising Rise to Power*. Holt.

Mikkelson, David. 2008. Receipt of Special 'Closed' Signs by Bank of America Signals that U.S. Banks Will Soon Be Shut Down by the Government for One Week. Snopes.com (October 6th).

Moore, Jim. 1990. *Conspiracy of One: The Definitive Book on the Kennedy Assassination*. Summit.

Munster, Anna. 2013. *An Aesthesia of Networks: Conjunctive Experience in Art and Technology*. MIT Press.

National Geographic. 2005. UFOs. *Is It Real?* Season One, Episode 2 (April 25th).

Nelson, Soraya Sarhaddi. 2014. Twenty Five Years After Death, A Dictator Still Casts a Shadow in Romania. *All Things Considered* (24th December). NPR.

Novella, Steven. 2018. *The Skeptic's Guide to the Universe: How to Know What's Really Real in a World Increasingly Full of Fake.* Grand Central.

Nyhan, Brendan, and Jason Reifler. 2010. When Corrections Fail: The Persistence of Political Misperceptions. *Political Behavior* 32:2.

Ochmann, Sophie, and Max Roser. 2019. Smallpox. OurWorldInData.org.

Olmsted, Kathryn S. 2019 [2009]. *Real Enemies: Conspiracy Theories and American Democracy, World War I to 9/11.* Oxford University Press.

Oreskes, Naomi, and Erik M. Conway. 2010. *Merchants of Doubt: How a Handful of Scientists Obscured the Truth from Tobacco Smoke to Global Warming.* Bloomsbury.

Orr, Martin, and Ginna Husting. 2007. Dangerous Machinery: 'Conspiracy Theorist' as a Transpersonal Strategy of Exclusion. *Symbolic Interaction* 30:2.

Ortega, Tony. 2008. The Phoenix Lights Explained (Again). eSkeptic (May 21st).

Osborn, Andrew. 2009. Adolf Hitler Suicide Story Questioned After Tests Reveal Skull Is a Woman's. *The Telegraph.*

Passas, Nikos. 1995. The Mirror of Global Evils: A Review Essay on the BCCI Affair. *Justice Quarterly* 12:2 (June).

Pelkmans, Mathijs, and Rhys Machold. 2011. Conspiracy Theories and Their Truth Trajectories. *Focaal: Journal of Global and Historical Anthropology* 59.

Pigden, Charles. 1995. Popper Revisited, or What Is Wrong with Conspiracy Theories? *Philosophy of the Social Sciences* 25:1.

Radford, Benjamin. 2010. How Were the Egyptian Pyramids Built? *Live Science* (June 1st) <www.livescience.com/32616-how-were-the-egyptian-pyramids-built-.html>.

Rao, T.S. Sathyanarayana, and Chittaranjan Andrade. 2011. The MMR Vaccine and Autism: Sensation, Refutation, and Fraud. *Indian Journal of Psychiatry.*

RationalWiki. 2019. Apophenia <https://rationalwiki.org/wiki/Apophenia>.

Redfern, Nick. 2015. *Secret History: Conspiracies from Ancient Aliens to the New World Order.* Visible Ink.

Rennie, John. 2009. Seven Answers to Climate Contrarian Nonsense. *Scientific American* (November 30th).

Said, Edward W. 1985. Orientalism Reconsidered. *Cultural Critique* 1:1.

Sargent, Mark K. 2015. *Flat Earth Clues.* YouTube <www.youtube.com/playlist?list=PLltxIX4B8_URNUzDE2sXctnU AEXgEDDGn>.

Schick, Theodore Jr., and Lewis Vaughn. 2019 [2001]. *How to Think about Weird Things: Critical Thinking for a New Age*. McGraw-Hill.

Schiff, Stacy. 2015. *The Witches: Suspicion, Betrayal, and Hysteria in 1692 Salem*. Little, Brown.

Schlafly, Phyllis. 2014 [1964]. *A Choice Not an Echo: Updated and Expanded 50th Anniversary Edition*. Regnery.

Shapira, Ian. 2018. Trump Delays Full Release of Some JFK Assassination Files until 2021. *Washington Post*.

Sheehan, Neil, et al., eds. 2017 [1971]. *The Pentagon Papers: The Secret History of the Vietnam War*. Racehorse.

Skeptical Science. 2016. The Cook et al. (2013) 97% Consensus Result Is Robust. <https://skepticalscience.com/97-percent-consensus-robust.htm>.

Solomon, Lawrence. 2008. *The Deniers: The World-Renowned Scientists Who Stood Up Against Global Warming Hysteria, Political Persecution, and Fraud*. Richard Vigilante.

Sports Illustrated. 1999. Michael Jordan Chronology. *Sports Illustrated* (January 12th).

Steyn, Mark, ed. 2015. *A Disgrace to the Profession: The World's Scientists—in Their Own Words—on Michael E. Mann, His Hockey Stick, and Their Damage to Science*. Stockade.

Strauss, Neil. 1994. Kurt Cobain's Downward Spiral: The Last Days of Nirvana's Leader. *Rolling Stone*.

Sunstein, Cass, and Adrian Vermeule. 2009. Conspiracy Theories: Causes and Cures. *Journal of Political Philosophy* 17:2.

Swift, A. 2013. Majority in U.S. Still Believe JFK Killed in a Conspiracy <www.gallup.com/poll/165893/majority-believe-jfk-killed-conspiracy.aspx>

Thomas, David. 1995. The Roswell Incident and Project Mogul. *CSI* (August 1st).

Thomson, Garrett, and Marshall Missner. 2000. *On Aristotle*. Wadsworth.

Thucydides. 2009. *The History of the Peloponnesian War*. Project Gutenberg <www.gutenberg.org/files/7142/7142-h/7142-h.htm>.

Townsend. Tim. 2013. Paranormal Activity: Do Catholics Belief in Ghosts? *U.S. Catholic* (October 30th).

Tran, Dari Sylvester. 2019. *Unrigging American Elections: Reform Past and Prologue*. Macmillan.

Uscinski, Joseph E. ed. 2018. *Conspiracy Theories and the People Who Believe Them*. Oxford University Press.

Van Prooijen, Jan-Willem. 2018. *The Psychology of Conspiracy Theories*. Routledge.

Van Prooijen, Jan-Willem, André P.M. Krouwel, and Thomas V. Pollet. 2015. Political Extremism Predicts Belief in Conspiracy Theories. *Social Psychological and Personality Science* 6:5.

Walters, Guy. 2010. *Hunting Evil: The Nazi War Criminals who Escaped and the Quest to Bring Them to Justice*. Broadway.

Ward, Charlotte, and David Voas. 2011. "The Emergence of Conspirituality." Journal of Contemporary Religion 26 (1):103–21. https://doi.org/10.1080/13537903.2011.539846;

Warner, Benjamin R., and Ryan Neville-Shepard. 2014. Echoes of a Conspiracy: Birthers, Truthers, and the Cultivation of Extremism. *Communication Quarterly* 62:1.

Warren Commission Report. 1992 [1964]. *The Warren Commission Report: The Official Report on the President's Commission on the Assassination of President John F. Kennedy*. Longmeadow.

Wayne, Gary. 2014. *The Genesis 6 Conspiracy: How Secret Societies and the Descendants of Giants Plan to Enslave Humankind*. Trusted.

Webber, Frances. 2000. Justice and the General: People vs. Pinochet. *Race and Class* 41:4.

Weinmann, Karlee, and Kim Bhasin. 2011. 14 False Advertising Scandals that Cost Brands Millions. *Business Insider* (September 16th).

Wood, Michael J. 2016. Some Dare Call It Conspiracy: Labeling Something a Conspiracy Theory Does Not Reduce Belief in It. *Political Psychology* 37:5.

Wood, Michael J., and Karen M. Douglas. 2013. 'What About Building 7?' A Social Psychological Study of Online Discussion of 9/11 Conspiracy Theories. *Frontiers in Psychology* 4:1–9.

Woodward, Bob. 2005. *The Secret Man: The Story of Watergate's Deep Throat*. Simon and Schuster.

Zia-Ebrahimi, Reza. 2018. When the Elders of Zion Relocated to Eurabia: Conspiratorial Racialization in Antisemitism and Islamophobia. *Patterns of Prejudice* 52:4.

The Illuminati

AMIN ASFARI is an Associate Professor of Criminal Justice at Wake Tech College. His interests include the study of Islamophobia and its relationship to antisemitism, as well as various criminological topics. His forthcoming edited volume is titled *Civility, Nonviolent Resistance, and the New Struggle for Social Justice.*

MARCO ANTONIO AZEVEDO is Adjunct Professor in Philosophy at University of Vale do Rio dos Sinos (Unisinos), Brazil. His main interests are in issues on metaethics, ethical theory, bioethics, and philosophy of medicine. He has published articles in the *Journal of Evaluation in Clinical Practice* and the *Journal of Medicine and Philosophy*. He wrote chapters for *Westworld and Philosophy* (2018), *The Good Wife and Philosophy* (2013) and *Bad Arguments: 100 of the Most Important Fallacies in Western Philosophy* (2019).

KENDAL BEAZER teaches microbiology in the Medical Laboratory Science department at Weber State University. He works part-time for Utah Public Health Laboratory on infectious diseases where he collaborates with other scientists and epidemiologists to prevent microbial outbreaks. Much of his research and curriculum is in vaccine-preventable diseases.

GREGORY L. BOCK is Assistant Professor of Philosophy and Religion at The University of Texas at Tyler. He is also Director of UT Tyler's Center for Ethics and the Philosophy, Religion, and Asian Studies programs. His research interests include forgiveness, bioethics, and the philosophy of religion. He is editor of *The Philosophy of Forgiveness, Volume III: Forgiveness in World Religions*, and *The Philosophy of Forgiveness, Volume IV: Christian Perspectives on Forgiveness.*

ROD CARVETH is an Associate Professor in Multimedia Journalism and Director of Graduate Studies for the School of Global Journalism and Communication at Morgan State University.

JEFF CERVANTEZ is an associate professor of philosophy and religious studies at Crafton Hills College.

BRETT COPPENGER is Assistant Professor of Philosophy at Tuskegee University. His philosophical interests include Epistemology, the Philosophy of Science, the Philosophy of Religion, as well as explaining the importance and relevance of philosophical ideas to popular-level audiences. His recent publications include editing (with Michael Bergmann) *Intellectual Assurance: Essays on Traditional Epistemic Internalism* (2016). He has also contributed chapters in *Westworld and Philosophy*, *The Man in the High Castle and Philosophy*, and *Arrested Development and Philosophy*.

M R.X. DENTITH is author of *The Philosophy of Conspiracy Theories* (2014) and *Understanding Conspiracy Theories: 9/11, QAnon and Beyond* (forthcoming), as well as the editor of *Taking Conspiracy Theories Seriously* (2018). Dentith is Aotearoa New Zealand's pre-eminent expert on Romanian conspiracy theories, and has been described by Joseph E. Uscinski as "one of the most important social epistemologists studying conspiracy theories." They are currently a Teaching Fellow in Philosophy at the University of Waikato.

DON FALLIS is a Professor of Philosophy at Northeastern University. He has written many philosophy articles on lying and deception, including "What is Lying?" in the *Journal of Philosophy* and "The Most Terrific Liar You Ever Saw in Your Life" in *The Catcher in the Rye and Philosophy*.

MICHAEL GOLDSBY is an associate professor of philosophy at Washington State University. He specializes in philosophy of science and has written articles on philosophical issues related to scientific modeling, randomized clinical trials, and climate change. He is also the lone philosopher among a team of engineers and scientists who are working on a project to preserve Food-Energy-Water security in the Pacific Northwest in the face of climate change.

RICHARD GREENE is a Professor of Philosophy at Weber State University. He formally served as Executive Director of the Intercollegiate Ethics Bowl. He is the author of *Spoiler Alert! (It's a Book About the Philosophy of Spoilers)*. He has co-edited a number of books on popular culture and philosophy including *The Princess Bride and Philosophy*, *Twin Peaks and Philosophy*, *Quentin Tarantino and Philosophy*, *Boardwalk Empire and Philosophy*, and *The Sopranos and Philosophy*. He is the co-host of the philosophy and pop culture podcast *I Think, Therefore I Fan*.

JOSHUA HETER is the co-editor of *The Man in the High Castle and Philosophy* and *Westworld and Philosophy*. He earned his PhD in Philosophy from Saint Louis University, and he is currently an Assistant Professor of Philosophy at Jefferson College in Hillsboro, Missouri.

RON HIRSCHBEIN initiated a program in war and peace studies at California State University, Chico. He has also held visiting professorships at University of California campuses in Berkeley and San Diego, and at the United Nations University in Austria. A frequently contributor to the philosophy and popular cultures series, he authored five books relating to issues of war and peace. He and Amin are expanding their current contribution into a book length manuscript.

ALEXANDER E. HOOKE is professor of philosophy at Stevenson University. He is co-editor of *The Twilight Zone and Philosophy* (Open Court, 2019), and author of *Philosophy Sketches—700 Words at a Time* (2018) and *Alphonso Lingis and Existential Genealogy* (2019).

MARK HUSTON is the Chair of Philosophy at Schoolcraft College in Livonia, Michigan. He has published in a variety of areas including philosophy of sport (tennis and golf), language, and film. He has also given many lectures, both academic and public, on conspiracy theories. Mark has recently published "Beyond 'Apocalypse Now': Just War Theory and Existentialism" in *The Community College Humanities Review* and "Medical Conspiracy Theories and Medical Errors" in the *International Journal of Applied Philosophy*.

DAVID KYLE JOHNSON is Professor of Philosophy at King's College, in Wilkes-Barre, Pennsylvania, and also produces lecture series for The Teaching Company's *The Great Courses*. His specializations include metaphysics, logic, and philosophy of religion and his "Great Courses" include *Sci-Phi: Science Fiction as Philosophy*, *The Big Questions of Philosophy*, and *Exploring Metaphysics*. Kyle is the editor-in-chief of *The Palgrave Handbook of Popular Culture as Philosophy*, and has edited volumes on popular culture and philosophy, including *Inception and Philosophy: Because It's Never Just a Dream* (2011) and *Black Mirror and Philosophy: Dark Reflections* (2019).

GARY JOHNSON is a Professor of Political Science at Weber State University. His work has been published in *State and Local Government Review*, *Public Administration Review*, and *Municipal Finance Review*. He co-authored *The Adapted City* (2004).

W. JOHN KOOLAGE is a professor of philosophy at Eastern Michigan University. He loves teaching and learning, and he is particularly fond of introducing people to philosophical thought. His research tends to focus on questions of scientific epistemology and cognition, especially as they relate to contemporary social issues.

DANIEL KRASNER is Associate Professor of Philosophy at Metropolitan State University of Denver, specializing in Philosophy of Language and Logic. He was first exposed to Leninism by close relatives.

COURTLAND LEWIS is Program Co-ordinator and Associate Professor of Philosophy and Religious Studies at Owensboro Community and Technical College. Specializing in Ethics, Forgiveness, and Justice, Court is the author of *Repentance and the Right to Forgiveness*, Series Editor of Vernon Press's series *The Philosophy of Forgiveness*, author of *Way of the Doctor: Doctor Who and the Good Life*, and editor and contributor of several other popular and academic publications.

PAUL LEWIS is Associate Professor and Chair of the Department of Philosophy at University of the Incarnate Word.

EDUARDO VICENTINI DE MEDEIROS is Adjunct Professor of Ethics at Universidade Federal de Santa Maria (UFSM), Brazil. His present research focuses mainly on counterfactual thinking in decision-making about moral values.

CHARLES PIGDEN is a graduate of King's College, Cambridge and did his doctorate as Commonwealth Scholar at La Trobe University, Melbourne, Australia. Since 1988 he has taught philosophy at the University of Otago, Dunedin, New Zealand, where for many years he was head of the Philosophy, Politics and Economics program. He is the editor of *Russell on Ethics* (1999), *Hume on Motivation and Virtue* (2009) and *Hume on Is and Ought* (2010). His interest in conspiracy theories dates back to his 1995 paper "Popper Revisited or What's Wrong with Conspiracy Theories" (to which question his answer is that although there is a lot wrong with *some* conspiracy theories, there is nothing wrong with conspiracy theories *as such*).

RACHEL ROBISON-GREENE is the editor or co-editor of thirteen books on philosophy and popular culture including, most recently, *The Handmaid's Tale and Philosophy*. She is a regular contributor to the ethics periodical *The Prindle Post*, where she writes frequently on topics pertaining to the environment, criminal justice reform, and ethics and technology. Rachel teaches philosophy at Utah State University. She is the co-host of the podcast *I Think, Therefore I Fan*.

JAMES ROCHA is an assistant professor of Philosophy at Fresno State. He has published on ethics, political philosophy, and philosophy of law. His forthcoming book is *The Ethics of Hooking Up: Casual Sex and Moral Philosophy on Campus* (Routledge).

MONA ROCHA is an instructor of Classics at Fresno State. She has published in history, women's studies, and cultural studies. Her forthcoming book is *Militant Feminism: The Women of the Weather Underground Organization*. She and James Rocha also have a forthcoming book: *Joss Whedon, Anarchist?*

BRENDAN SHEA is an Instructor of Philosophy at Rochester Community and Technical College and a Resident Fellow at the

Minnesota Center for Philosophy of Science. His teaching and research focus mainly on applied ethics, the philosophy of science, and the areas where these intersect. He has also published a number of book chapters on popular culture and philosophy, on topics ranging from *Alice in Wonderland* to *Downton Abbey* to *The Princess Bride*.

Index

Printed in the USA
CPSIA information can be obtained
at www.ICGtesting.com
JSHW012023140824
68134JS00033B/2848

9 780812 694796